Quality of Service in Wireless Networks Over Unlicensed Spectrum

Synthesis Lectures on Mobile and Pervasive Computing

Editor

Mahadev Satyanarayanan, Carnegie Mellon University

Mobile computing and pervasive computing represent major evolutionary steps in distributed systems, a line of research and development that dates back to the mid-1970s. Although many basic principles of distributed system design continue to apply, four key constraints of mobility have forced the development of specialized techniques. These include unpredictable variation in network quality, lowered trust and robustness of mobile elements, limitations on local resources imposed by weight and size constraints, and concern for battery power consumption. Beyond mobile computing lies pervasive (or ubiquitous) computing, whose essence is the creation of environments saturated with computing and communication yet gracefully integrated with human users. A rich collection of topics lies at the intersections of mobile and pervasive computing with many other areas of computer science.

Quality of Service in Wireless Networks Over Unlicensed Spectrum
Klara Nahrstedt

The Landscape of Pervasive Computing Standards
Sumi Helal

A Practical Guide to Testing Wireless Smartphone Applications
Julian Harty

Location Systems: An Introduction to the Technology Behind Location Awareness
Anthony LaMarca and Eyal de Lara

Replicated Data Management for Mobile Computing
Douglas B. Terry

Application Design for Wearable Computing
Dan Siewiorek, Asim Smailagic, and Thad Starner

Controlling Energy Demand in Mobile Computing Systems
Carla Schlatter Ellis

RFID Explained
Roy Want

Quality of Service in Wireless Networks Over Unlicensed Spectrum
Klara Nahrstedt

ISBN: 978-3-031-01354-6 print
ISBN: 978-3-031-02482-5 ebook

DOI 10.1007/978-3-031-02482-5

A Publication in the Springer series
SYNTHESIS LECTURES ON MOBILE AND PERVASIVE COMPUTING

Lecture #8

Series Editor: Mahadev Satyanarayanan, Carnegie Mellon University

Series ISSN

ISSN 1933-9011 print

ISSN 1933-902X electronic

Quality of Service in Wireless Networks Over Unlicensed Spectrum

Klara Nahrstedt
University of Illinois at Urbana-Champaign

SYNTHESIS LECTURES ON MOBILE AND PERVASIVE COMPUTING # 8

ABSTRACT

This Synthesis Lecture presents a discussion of Quality of Service (QoS) in wireless networks over unlicensed spectrum. The topic is presented from the point of view of protocols for wireless networks (e.g., 802.11) rather than the physical layer point of view usually discussed for cellular networks in the licensed wireless spectrum. A large number of mobile multimedia wireless applications are being deployed over WiFi (IEEE 802.11) and Bluetooth wireless networks and the number will increase in the future as more phones, tablets, and laptops are equipped with these unlicensed spectrum wireless interfaces.

Achieving QoS objectives in wireless networks is challenging due to limited wireless resources, wireless nodes interference, wireless shared media, node mobility, and diverse topologies. The author presents the QoS problem as (1) an optimization problem with different constraints coming from the interference, mobility, and wireless resource constraints and (2) an algorithmic problem with fundamental algorithmic functions within wireless resource management and protocols.

KEYWORDS

quality of Service in 802.11 wireless networks, bandwidth management, delay management, price-based rate allocation, cross-layer QoS framework, delay adaptation, dynamic soft-real-time scheduling, location-based routing, fault-tolerant routing, energy-efficient routing

Contents

Preface xiii

1. Basics of Quality of Service in Wireless Networks 1
- 1.1 What Is Quality of Service? 1
- 1.2 Wireless Network Characteristics 3
  - 1.2.1 Single-Hop Wireless Local Area Network Characteristics 6
  - 1.2.2 Multi-Hop Wireless Ad Hoc Network Characteristics 9
- 1.3 QoS Modeling 10
  - 1.3.1 Deterministic QoS 11
  - 1.3.2 Proportional QoS 11
  - 1.3.3 Statistical QoS 12
- 1.4 Functional Components Toward QoS Provisioning 13

2. QoS-Aware Resource Allocation 17
- 2.1 Introduction 17
- 2.2 System Model 19
  - 2.2.1 Shared Wireless Channel 20
  - 2.2.2 Contention Model and Resource Constraints 21
  - 2.2.3 Integrating Example 24
- 2.3 Resource Allocation Problem Formulations 27
  - 2.3.1 Utility-Based Resource Allocation Problem 28
  - 2.3.2 Price-Based Resource Allocation Problem 29
- 2.4 Practical Issues 31
- 2.5 Summary 33

3. Bandwidth Management 35
- 3.1 Introduction 35
- 3.2 Price-Based Algorithm for Rate Allocation 36
  - 3.2.1 Per-Clique Price and Flow Rate Calculation 40
  - 3.2.2 Decentralized Clique Construction 41

3.2.3 Two-Tier Algorithm Integration 43
3.2.4 Practical Issues 45
3.3 Dynamic Bandwidth Management 49
3.3.1 Bandwidth Management Protocol 52
3.3.2 Total Bandwidth Estimation 56
3.3.3 Bandwidth Allocation and Adaptation 58
3.3.4 Practical Issues 62
3.4 Summary 66

4. Delay Management 69
4.1 Introduction 69
4.2 Delay Control with Upper Layers Adaptation 71
4.2.1 Cross-Layer QoS Framework 73
4.2.2 Proportional Delay Differentiation Scheduler 75
4.2.3 Adaptors Design 81
4.2.4 Practical Issues 87
4.3 Integrated Dynamic Soft Real-Time Framework 92
4.3.1 iDSRT Architecture 94
4.3.2 Scheduling Components 96
4.3.3 Deadline Assignment Algorithm 103
4.3.4 Practical Issues 104
4.4 Summary 106

5. Routing 109
5.1 Introduction 109
5.2 Predictive Location-Based QoS Routing 111
5.2.1 Location Prediction 112
5.2.2 QoS Routing 116
5.2.3 Practical Issues 117
5.3 Fault-Tolerant Routing 121
5.3.1 Routing Algorithm Overview 124
5.3.2 Finding Feasible Paths 125
5.3.3 Routing Protocol Design 128
5.3.4 Practical Issues 130
5.4 Energy-Efficient Distributed Routing 134
5.4.1 Network and Traffic Demand Models 137

5.4.2 Power Consumption Model ... 138
5.4.3 Maximum Network Lifetime Routing Problem ... 140
5.4.4 Distributed Routing ... 142
5.4.5 Practical Issues ... 145
5.5 Summary ... 147

Acknowledgment ... 149

References ... 151

Author Biography ... 161

Preface

This lecture presents a discussion of Quality of Service (QoS) in wireless networks over unlicensed spectrum. The reason for this lecture is that the licensed spectrum is getting saturated due to the pervasiveness of handheld devices and users' rising demand for multimedia content, such as movies and television content, to be delivered on handheld devices. Hence, we are seeing and will see more and more multimedia traffic going over unlicensed spectrum with protocols such as IEEE 802.11. However, to yield high quality in multimedia content transmission and delivery between content providers and consumers, one needs to pay attention to Quality of Service, similarly as is done in network protocols and resource management over licensed spectrum. This lecture aims to provide the entry point for interested readers such as engineers and researchers within service providers/companies who develop and/or provide multimedia services to mobile customers or graduate students and to other researchers who are conducting studies of the next generation of wireless networks over unlicensed spectrum for new mobile multimedia applications.

The lecture's aim is not to be comprehensive, since the QoS topic in wireless networks and wireless networks over unlicensed spectrum has a very broad scope, and it has been explored in the research community and by the industry at large over the past 15–20 years. The lecture's goal is to provide insights into some of the problems when one embarks on designing QoS in wireless networks over unlicensed spectrum for multimedia applications. The focus is on some of the crucial QoS-aware functions, abstractions, models, and corresponding protocols in this space, as well as lessons learned that I have gathered over the past 10 years working in this area.

Chapter 1 introduces the QoS concept, the scope of the types of wireless networks, and the QoS-aware functions we consider throughout the lecture. Chapter 2 is about formulating QoS and resource allocation problems as optimization problems. Chapter 3 discusses QoS-aware functions, abstractions, models, algorithms, and protocols for bandwidth allocation and management, and Chapter 4 is about delay control and management functions and their corresponding abstractions, algorithms, and protocols. Chapter 5 is about QoS-aware routing, and it completes the key set of functions to consider when embedding QoS capabilities within wireless networks over unlicensed spectrum.

Chapter 1 has the main goal of introducing readers to the overall concept of Quality of Service (QoS) as well as identifying the network scope where the QoS concept will be considered, such as the single hop wireless networks and mobile wireless networks over unlicensed spectrum. The chapter also outlines the key set of functions that need to be considered in order to gain certain levels of QoS such as admission control, resource allocation, monitoring, scheduling, and routing.

Chapter 2 introduces readers to the resource allocation problem in wireless networks over unlicensed spectrum. Important concepts, such as interference among nodes, interference/contention graphs, clique concepts in contention graphs, and utility-based problems and price-based problem formulations are discussed. The aim is to have a reader look at the provisioning of QoS as a resource optimization problem and show the strong differences when provisioning QoS in wired and wireless networks.

Chapter 3 concentrates on bandwidth as the network resource and the goal is to provide rate/bandwidth guarantees over unlicensed spectrum networks. This chapter presents bandwidth allocation, enforcement, and management functions with their algorithmic designs, which represent theoretical and practical solutions to the problems articulated in Chapter 2.

Chapter 4 is about delay guarantees, and the goal is to introduce delay functions with their algorithmic designs. We present delay functions that enforce delay guarantees such as allocation, monitoring and scheduling, and adaptation functions that respond to varying demands coming from users and applications, and underlying hardware, system, and network constraints.

Chapter 5 discusses routing functions that aim to preserve QoS guarantees as traffic is routed through a wireless network. The goal is to introduce readers to a large diversity of considerations when routing occurs since different contextual information may influence and assist in routing. This chapter presents routing difficulties, algorithmic solutions, and protocols, if one considers important contexts such as location, reliability, and energy that guide routing in wireless networks over unlicensed spectrum.

Each chapter finishes with its own conclusion discussion; hence, the overall lecture concludes with a large number of references which might be helpful to readers as they dive into this important and exciting area of broad interest.

Klara Nahrstedt

CHAPTER 1

Basics of Quality of Service in Wireless Networks

1.1 WHAT IS QUALITY OF SERVICE?

The overall lecture aims to discuss diverse issues related to quality of service (QoS) in wireless networks over unlicensed spectrum. However, before we attempt to understand issues of QoS in wireless networks, it is important to focus on what is understood by quality and the meaning implied by quality in the context of QoS. There are multiple definitions of quality in the economics literature. For example, the British Standards Institution [81] defines quality as "the totality of features and characteristics of a product or service that bear on its ability to satisfy stated or implied needs." Another source [82] defines quality as "quality refers to the amounts of the un-priced attributes contained within each unit of a priced product." Most of the definitions incorporate the concept of "characteristics" and consider quality in the context of a particular product. It means quality is referred to in terms of characteristics or attributes associated with a product or service, where things can be measured. However, one needs to understand that *quality means more than one can easily measure*. For example, Crosby [84] defines quality simply as "conformance to requirements," taking into account customer's perception.

In telecommunication and networked systems like wireless networks, the product/service is the *network service*, and its service levels in terms of *connectivity and performance* (measured in network association setup time, data throughput, data end-to-end delay, and loss of information) very much influence the user (customer) in its usage, productivity, and satisfaction. For example, if a user cannot connect her laptop wirelessly to the Internet at a business meeting to show a business-relevant video clip from a remote video server, she becomes dissatisfied with the network service. If she finally connects, but she gets very low data throughput between her laptop and Internet for the video streaming task, she will get truly frustrated, since after finally getting connected (possibly after several trials), she cannot use the wireless channel for her video task, and her productivity might suffer in order to work over a slow connection. In this case, she rather drops the wireless connection and performs alternative tasks during her business meeting. Notice in the above example the major tension between the minimal network service level of "providing connectivity" only and the desired network service level of "providing productive throughput" for a user in addition to the connectivity. In the modern networks, we often go with the basic notion of making the service available as fast as possible through statistical multiplexing without any end-to-end guarantees. This approach is often beneficial to the service provider because many customers can be served. However, as the above example shows, the user/customer wants not only the fast connectivity of the wireless laptop to the Internet, but also a certain throughput guarantee. If the connectivity does not come with a certain end-to-end guarantee of a "productive throughput" (e.g., 5–10 Mbps for the lifetime of the session), the connectivity becomes meaningless to accomplish the desired business task.

The above network service QoS discussion goes along the arguments of QoS for telecommunication services, presented in Reference [85]. Quality of service must be considered in the context of the customer/user and in the context of the network provider. In the context of the *user*, QoS is defined by the attributes which are considered essential in the *use of the service*. In the context of the *network service provider*, QoS is defined by parameters which contribute toward the *end-to-end performance of the service*, where the end-to-end performance must reflect the user's requirements. Note that the QoS definition discussion goes very much along the economists work of Lancaster [86] (the concept of the attributes that are of benefit to the customer) and Crosby [84] (conformance to customer requirements throughout the entire product/service life cycle). Throughout the lecture, when discussing QoS, we will talk about characteristics of networked services and their QoS as follows:

- quality parameters should provide benefits to the user and to the network service provider;
- quality is expressed on a service-by-service basis (Note that since networks comply with layered designs ranging from application, middleware (session), transport, network, medium access (MAC) to physical layers, each service within each layer may express its own quality.)
- quality parameters might be different from the customer's or service provider's viewpoint;
- for the customer, quality parameters are expressed on an end-to-end basis.

It is important to clarify that often network performance has been mistaken for *QoS*. Network performance is the technical performance of network elements or of the whole network, where quality is experienced by the user and user/customers state what levels of performance are required. Hence, network performance contributes toward QoS and should be derived from QoS to be offered to the customers and incorporated into network schedules, traffic shaping, and other network functions within layers to achieve the desired QoS.

In summary, quality of service and end-to-end guarantees are and will be very important for successful service delivery to the customers and for the acceptance of new multimedia services as

TABLE 1.1: Notations for Chapter 1

NOTATION	DEFINITION
DIFS	Distributed Interframe Space
PIFS	Point Interframe Space
SIFS	Short Interframe Space
b	Back-off time
[0,CW]	Contention window (in time slots)
$\mathrm{CW}_{\mathrm{min}}$	Minimum size of CW
$\mathrm{CW}_{\mathrm{max}}$	Maximum size of CW
$\mathrm{CW} = 2^B C_{\mathrm{min}}$	Content window with B maximum number of backoffs
$f \in F$	Flow f from set of flows F
$B_{\mathrm{min}}(f)$	Minimum bandwidth required for flow f
$B_{\mathrm{max}}(f)$	Maximum bandwidth required for flow f
d_{min}	Minimum required delay for flow f
d_{max}	Maximum required delay for flow f
$d, d_1, d_2, \ldots, d_N$	Delay and delay levels from 1 to N
RTT	Round-trip time
C	Number of service classes

they rapidly grow in our economy. This also means that the tension between the service providers and customers in terms of servicing everybody with best-effort services vs servicing customers according to their end-to-end QoS requirements will grow even more. The reason is that the service providers will not be able for long to fall back to overprovisioning utilizing the Moore's law in order to provide everybody satisfactory network performance and, hence, quality of service to the users. We are hitting limits of the Moore's law, and the data traffic demands are increasing every day due to increased multimedia traffic demands from the YouTube, Flicker, Skype, and other multimedia-providing services. Users are demanding high quality in data delivery from their service provider, and if not, they move to alternative competing service provider(s) to get their demands satisfied.

In this chapter, we want to concentrate on the *network performance*, *QoS* and *end-to-end QoS guarantees* in wireless networks, using unlicensed spectrum. In Section 1.2, we will discuss characteristics of these wireless network and stress that wireless resource management and the various issues of QoS very much depend on the wireless network topology, mobility, flows, and other network characteristics. In Section 1.3, we will discuss various QoS models that might be considered in IEEE 802.11-like wireless networks, and we conclude the chapter with Section 1.4, discussing the main resource management operations and fundamental components that will bring the wireless network performance toward desired QoS levels. Throughout the chapter, we will use notation and definitions as shown in Table 1.1.

1.2 WIRELESS NETWORK CHARACTERISTICS

In this lecture, we assume wireless devices such as phones, sensors, laptops, tablets, vehicles, which have become pervasive. Their fast spread has enabled the design and wide deployment of not only *cellular wireless networks*, using *licensed spectrum*, but also wireless networks, using unlicensed spectrum, such as WiFi and Bluetooth. In this section, we discuss first the *concept of the unlicensed spectrum* and the assumed wireless network characteristics over unlicensed spectrum. Second, we discuss briefly *delay and bandwidth-sensitive applications*, and third, we discuss *different topologies* of the assumed wireless networks. Note that after this section, we will use the terms of "wireless networks" and "wireless networks over unlicensed spectrum" synonymously.

Unlicensed spectrum: Unlicensed spectrum, or *license-free spectrum* as it is sometimes called, simply means a spectrum band that has rules predefined for both the hardware and deployment methods of the radio in such a manner that interference is mitigated by the technical rules (e.g., IEEE 802.XX) defined for the bands rather than it being restricted for use by only one entity through a spectrum licensing approach. Any person or entity that does not infringe upon the rules for the equipment or its use can put up a license-free network at any time for either private or public purposes including commercial high speed Internet services. Some of the most commonly used license-free frequencies in the United States are at 900 MHz, 2.4 GHz, 5.2/5.3/5.8 GHz, 24 GHz, and above 60 GHz. There are other bands such as the band at 4.9 GHz which is allocated for public safety use. The rules vary from band to band (e.g., IEEE 802.11 rules are different than IEEE 802.15 rules). The rules for an equipment in each band vary somewhat as do the power allotment and configuration of the equipment.

One example of license-free service is the *WiFi-enabled wireless service*. The spectrum used for these WiFi-enabled wireless networks is mostly at 2.4 and 5.2/5.3 GHz and the distance coverage range is between 20–70 m indoors and 100–250 m outdoors [87, 122]. For example, in the

United States, the 5.2/5.3 GHz band is used for both IEEE 802.11a hotspot access as well as outdoor use. The IEEE 802.11b uses the 2.4 GHz band. The IEEE 802.11 wireless standard solutions 802.11a/b/g/n are collectively known as WiFi technologies.

Another example of license-free service is the *Bluetooth technology* (belongs to the IEEE 802.15 standard family) that followed a different development path than the 802.11 family. Bluetooth supports a very short range (approximately 10 m) and relatively low bandwidth (1–3 Mbps in practice). It is designed for low-power network devices like handhelds [87, 122].

The third example for license-free service is the *Worldwide Interoperability for Microwave Access (WiMax) technology*, designed for long-range networks (spanning miles) as opposed to local area wireless networks. It is based on the IEEE 802.16 WAN (wide area network) communications standard family serving in bands ranging anywhere from 2 GHz up to 66 GHz [87, 122].

Delay and bandwidth-sensitive applications: Typical applications for these wireless networks include delay/throughput-sensitive as well as best effort applications running on mobile devices. Especially, the delay/throughput-sensitive *multimedia applications*, such as streaming multimedia content distribution (e.g., YouTube service), VoIP (voice over IP) (e.g., Skype service), real-time monitoring, and control of cyber-physical environments in power grid and health care, would very much benefit from QoS considerations in wireless networks. They benefit from wireless networks that know how to deal with Quality of Service (QoS) constraints such as bandwidth, delay, and packet loss (delivery ratio) since they need certain level of QoS guarantees to be adequate and yield effective communication. On the other hand, best effort applications such as file transfer are more tolerant to bandwidth and delay variations. To support coexistence of both types of applications in WiFi wireless networks, *QoS management*, *QoS-aware resource management*, and *QoS-aware protocols* must be used to keep appropriate state information, allocate resources to flows, and provide guarantees to delay/througput-sensitive traffic in the presence of best effort traffic. However, to efficiently design appropriate QoS management, and provide appropriate QoS-aware resource management/protocols, one needs to understand the characteristics of wireless networks.

Types of wireless networks: Wireless networks over unlicensed spectrum have very unique characteristics (very different from wireline networks and licensed cellular networks), and they impose great challenges on the design of QoS management and QoS-aware protocols/resource management. These unique characteristics very much depend on the types of wireless networks. In this series of chapters, we will consider four types of wireless networks: (a) single-hop static wireless LAN (Local Area Network) (static WLAN), (b) single-hop mobile wireless LAN (mobile WLAN), (c) static multi-hop wireless ad hoc network (static WANET), and (d) mobile multi-hop wireless ad hoc network (MANET). An example of a static WLAN is a single hop *power-grid sensor monitoring* network collecting sensory information to a collecting gateway. An example of mobile WLAN is the *network of laptops within a building*, communicating wirelessly with Access Points (APs) and through APs with each other. APs in the building ensure full wireless coverage and are connected via Internet with each other. An example of static multi-hop wireless networks is an *environmental sensory network* that collects sensory information from remote regions through multi-hop wireless network to a data sink. An example of mobile multi-hop wireless networks is a *network of vehicles or pedestrians* organized in a peer-to-peer ad hoc fashion that disseminates content from senders to receivers through multi-hop wireless network.

1.2.1 Single-Hop Wireless Local Area Network Characteristics

Single-hop WLANs over unlicensed spectrum often deploy the WiFi wireless network technology, running the IEEE 802.11 MAC layer protocol [1]. However, several other WLAN technologies over unlicensed spectrum, belonging to the IEEE 802.15 standard, are being deployed and starting to be widely used for personal area networks (PANs). Examples are the Bluetooth technology to provide a universal short-range wireless capability [87], and ZigBee technology, running the IEEE 802.15-4 standard and working at low data rate over short distances compared to WiFi, hence very suitable for low rate wireless personal area networks. Another example is WiMAX, an IP-based wireless broadband access technology and a wireless digital communication system, also known as IEEE 802.16, that is intended for wireless "metropolitan area networks." WiMAX provides performance similar to 802.11/Wi-Fi networks with the coverage and also QoS capabilities of cellular networks. WiMAX is a standard initiative and will be very good potential technology for QoS provisioning in unlicensed spectrum. The more recent Long Term Evolution (LTE) standard is a similar term describing a parallel technology to WiMAX that is being developed by vendors and carriers as a counterpoint to WiMAX [122].

Throughout the series of chapters, we will consider the IEEE 802.11 wireless LAN network and its IEEE 802.11 MAC layer protocols [1] due to their wide deployment, but some of the higher layer QoS solutions over WLAN will be also applicable if other MAC technologies for unlicensed spectrum are deployed (e.g., in case of sensor networks and Bluetooth networks).

Now, we discuss briefly the IEEE 802.11 MAC layer and its two different access modes, namely, *distributed coordination function (DCF)* and *point coordination function (PCF)* to set the context for future QoS understanding and assumptions.

IEEE 802.11 DCF [87] is the basic access mechanism. It uses a carrier sense multiple access with collision avoidance (CSMA/CA) algorithm to coordinate the medium access. Before a packet is sent, the node senses the medium. If it is idle for at least a distributed interframe space (DIFS) period of time, the packet is transmitted. Otherwise, a backoff time b (in time slots) is uniformly chosen from the range (0, CW), where CW is called the contention window. Initially, CW is set to the value of minimum contention window CW_{min}. After the medium has been detected idle for at least a DIFS time period, the backoff timer is decremented by one for each time slot during which the medium remains idle. If the medium becomes busy during the backoff process, the backoff timer is paused and is resumed when the medium has been sensed idle for a *DIFS* again. When the backoff timer reaches zero, the packet is transmitted. Upon detection of a collision, the *CW* is doubled, up to a maximum value $CW_{max} = 2^B CW_{min}$, where B is the maximum number of backoff stage. A new backoff time is then chosen, and the backoff procedure starts again. After a successful transmission, the CW is reset to CW_{min}. The backoff mechanism is also used after a successful transmission before sending the next packet. IEEE 802.11 DCF can be deployed on both centralized and distributed WLANs. Note that IEEE 802.11 DCF is mostly deployed in the current WiFi 802.11 wireless network infrastructures.

IEEE 802.11 PCF [87] is an alternative access method implemented on top of the DCF, and it is designed for the centralized wireless LAN model. It provides a contention-free polling-based medium access method and supports time-bounded applications. It means the operation consists of polling by a centralized polling master (point coordinator). The point coordinator makes use of PIFS (point intraframe space) time period when issuing polls. Because PIFS is smaller than DIFS, the point coordinator can seize the medium and lock out all asynchronous traffic, while it issues polls and receives responses.

One recognizes that this PCF mode of 802.11 network would be excellent for multimedia traffic and provide high QoS for time-sensitive traffic. However, it is not as optimal for a mixed traffic between time-sensitive and best effort (asynchronous) traffic. To illustrate an extreme situation of PCF mode of operation, let us consider the following possible scenario. A wireless network is configured so that a number of wireless stations with time-sensitive traffic are controlled by the point coordinator while remaining traffic contends for access using CSMA. The point coordinator could issue polls in a round-robin fashion to all stations configured for polling. When a poll is issued, the polled station may respond using short intraframe space time period, used for all immediate response actions. If the point coordinator receives a response, it issues another poll using PIFS. If no response is received during the expected turnaround time, the coordinator issues a poll. If this extreme situation would be implemented, the point coordinator would lock out all asynchronous traffic by repeatedly issuing polls.

To prevent this scenario, the PCF defines an interval, called the superframe. During the first part of this interval, the point coordinator issues polls in a round-robin fashion to all stations configured for polling. The point coordinator then idles for the remainder of the superframe, allowing a contention period for asynchronous access. However, overall, the PCF mode of operation is not very efficient for a mixed traffic, and it can allow for starving the best effort traffic. As we argued above, we want to support QoS for delay-throughput-sensitive applications, but also have a strong coexistence with best effort asynchronous applications over wireless networks, since otherwise, users might reject and not use the overall wireless network service. Hence, in this series of chapters, we will consider IEEE 802.11 DCF-based WLAN due to its wide deployment (IEEE 802.11 b/g versions).

Note that there is a specific *IEEE.802.11e* standard for support of QoS in WLANs [88]; however, it is not being widely deployed in current WiFi interfaces (mostly, we see in devices the support of 802.11a/b/g variants). For completeness, we will briefly introduce two modes of this standard, since we will see some of the concepts implemented in higher layers of the wireless protocol stack to delivery QoS. The two modes are the *Enhanced Distributed Coordination Function (EDCF)* and the *Hybrid Co-ordination Function (HCF)* modes.

The *EDCF mode* is an improved version of the DCF mode to provide QoS via differentiated service. QoS support is realized with the introduction of Traffic Categories (TC). Upon finding the channel idle or after collision occurs, higher priority TCs are likely to wait a shorter interval before attempting to transmit, while lower priority TCs wait longer.

The *HCF mode* is an improvement of the PCF mode. It supports both a contention-free period and a contention period. During the contention-free period, channel access is solely through polling from the Hybrid Coordinator (HC). During the contention period, channel access can be via listen-before-talk EDCF distributed coordination as well as through polling, when the HC's poll pulse wins the contention. The transition between the two periods is signaled using a beacon. A special time interval is set aside for mobile hosts to send the HC their resource requests, which the HC uses in determining polling frequency and length of transmit opportunities for the respective hosts, so that their requests can be satisfied. This solves the problem of unknown transmission times of polled stations in legacy IEEE 802.11 PCF. The IEEE 802.11e EDCF could prove useful to bandwidth management architectures (we will discuss bandwidth QoS and the corresponding bandwidth management architecture in Chapter 3), which need packet classification and differentiation. In terms of IEEE 802.11e HCF, it remains seen whether vendors implement this mode. As we already discussed, the legacy PCF mode has problems with support of asynchronous traffic, and with beacon delays that hinder its implementation. The same problems persist with HCF, and the current WiFi interfaces on laptops, mobile phones, and tablets are not delivering the 802.11e option.

In the WLAN environment, we can observe wireless nodes that are stationary and connect with other nodes via *AP* (*access point*) as well as mobile nodes that move and connect with each other either via AP or directly (ad hoc) with other nodes as shown in Figure 1.1. The stationary WLANs are usually characterized by (a) using AP; (b) having centralized control of resources above MAC layer (many QoS solutions build a modified PCF functionality at higher protocol layers over 802.11 DCF that aim to remedy some of the problems of PCF; we will show one such centralized approach [30] for achieving QoS in Chapter 4); (c) shared nature of wireless channel which needs

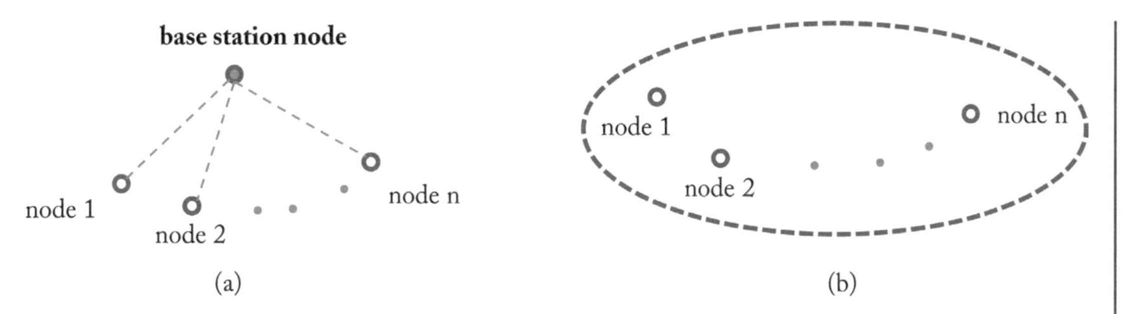

FIGURE 1.1: Models of (a) WLAN with AP and (b) ad hoc WLAN [2].

to be taken into account for QoS-aware resource allocation schemes, and (d) highly unpredictable and unreliable wireless channels where their capacities can vary. Examples of static WLANs are static wireless measurement devices (e.g., Phasor Measurement Units) in Smart Grid substations that connect over WLAN to a concentrator gateway or neighborhood smart meters that connect to a Smart Grid collection gateway [30]. *Mobile WLAN* are prevelant in our daily lifes as we see them in our offices, campuses, cities, and homes. Mostly, organizations install multiple WiFi APs and mobile users with their WiFi-enabled mobile devices, get access to the Internet via wireless connectivity, and association between the mobile device and the nearest WiFi AP. In the recent mobile phones (e.g., Android phones) and laptops, users can connect to each other in WiFi ad hoc mode (without AP) and exchange information with each other. Bluetooth technology also allows mobile WLAN configurations and exchange of information. Mobile WLAN is characterized as follows: (a) it can use APs or ad hoc modes; (b) it deploys centralized and distributed QoS-aware resource management above MAC layer. In the centralized model, a central coordinator coordinates packet transmission and performance of network tasks. In the distributed model, every mobile node plays the same role and carries the same responsibilities in resource allocation and network management; (c) it must consider the shared nature of wireless channels; (d) it must deal with mobility of nodes since mobility causes breaking connections; (e) it must consider channel unreliability and high volatility in resource availability.

1.2.2 Multi-Hop Wireless Ad Hoc Network Characteristics

Multi-hop Wireless Ad Hoc Networks (WANET) are less utilized in offices, campuses, homes and cities due to the large density of people and, hence, low cost/high density of APs deployment to provide extensive WiFi coverage to users and therefore wide wireless accessibility to Internet. However, in unstructured environments that occur after disasters or in military situtations, first responders or military personel will very much need WANET capabilities. Also, in remote agricultural fields, transportation rails, power lines, and forrest areas, we may desire to observe the health of animals, crop status, and other objects via sensors which then need for cost reasons to employ WANETs to collect status information about the observed objects. We will again classify WANETs into stationary WANETs and mobile WANETs, also called MANETs.

Characteristics of *stationary WANET* are mostly prevelant in sensor networks, where each sensor is fixed to a particular location, collecting an information such as temperature, humidity, survellaince images, and reporting the sensory information over wireless spectrum to a neighboring sensory device with the goal of forwarding the sensed information over multiple hopes toward a collecting aggregator node. This is especially useful in agricultural observations, animal behaviors identification, and other environmental applications. The issues to consider in these types of networks are (a) the shared nature of the wireless medium of sensor nodes since the neighboring nodes may interfer with each other transmission, (b) energy issues, since the sensors are usually very simple devices placed in remote locations, and battery replacement can be a difficult task, and (c) wireless channels can be highly unpredictable as various physical objects may cause disruptions (e.g., animals, snow).

The *MANETs* can be mostly found in first responder applications and in military scenarios where a structured AP infrastructure does not exist or was destroyed by a disaster (earthquake, hurricane, fire). However, it will be interesting to see if MANETs will find usage also in cities and campuses for social networking applications as mobile phone are being enabled with IEEE 802.11 ad hoc modes. The characteristics of MANET are as follows: (1) these networks have no centralized control, hence only local state information is available to any node in the network. QoS-aware protocols and algorithms must use distributed algorithms and not rely on global information; (2) the shared nature of the wireless channel makes resource allocation very complex since allocation of resources at an individual node affects available resources at its contending nodes, which might be outside of its communication range [3]; (3) mobility of nodes often breaks connectivity among nodes or resource association and allocation at each node, incurring high message overhead from the need to reconfigure the QoS and resource associations; (4) wireless channels are highly unpredictable, and their capacities can change dramatically. Hence, dealing with packet loss must be an inherent property of QoS-aware protocols.

1.3 QOS MODELING

Owing to limited wireless resources and other challenging characteristics of WiFi-based wireless networks, as discussed in Section 1.2, providing Quality of Service and control network performance toward achieving desired QoS requirements is very hard and definitely much harder than in wired networks. To understand the hardness of the QoS problem in wireless networks, we need to discuss the QoS models and their representations that the users/customers may use in their QoS requirements specification, and wireless service providers aim to conform to the QoS requirements and provide the network performance toward the desired QoS. Throughout the series of chapters, we will refer to three types of QoS models and their representations that one may find in wireless networks: (1) *deterministic QoS* model, (2) *proportional QoS* model, and (3) *statistical QoS* model.

1.3.1 Deterministic QoS

Deterministic QoS model in wireless networks mostly represents a conceptual model to express that a user desires exactly certain QoS level of delay or throughput, e.g., 80 ms end-to-end delay for voice communication or 10 Mbps bandwidth for video streaming during the entire wireless session. It is important to stress that in real wireless systems, we do not find deterministic QoS models implemented over contention-based wireless links such as IEEE 802.11, since it is impossible to deliver deterministic end-to-end QoS guarantees. One can come close to certain deterministic QoS guarantees, but the implementation is very expensive and complex. Hence, the current QoS models, service providers contemplate and provide, are mostly statistical QoS models.

One can see often usage of deterministic QoS specifications when expressing *acceptable QoS contraints, minimum and maximum performance bounds, and thresholds* against which QoS-aware resource operations check their performance. For example, if we consider a wireless network with flows $f \in F$, then the application initiating flow f can have a very specific and deterministic minimum bandwidth requirement $B_{\min}(f)$ and a maximum bandwdith requirement $B_{\max}(f)$. Another example can be the minimal and maximal delay constraints $d_{\min}$ and $d_{\max}$ of flow f's packets, and round-trip delay measurements of packets $(d_1, d_2, \ldots, d_N)$. These constraints, bounds, and thresholds are then used in various resource management frameworks, functions, services, and protocols by wireless service providers to come close to desired QoS requirements specified by users.

1.3.2 Proportional QoS

There is a growing need to provide service differentiation in wireless networks. It means to support network QoS, service differentiation divides network traffic into classes with different priorities, from which applications can choose to meet their QoS requirements. Existing studies on service differentiation in wireless networks have focused on the design of MAC layer protocols to achieve *distributed priority scheduling* [e.g., 25, 26, 27, 89]. While these works have shown to achieve certain differentiation (higher throughput or lower delay for higher priority packets) at the MAC layer, they did not consider the problem of service differentiation from the *end-to-end perspective*, which is especially important for multimedia applications. To solve this problem, one needs to introduce proportional service differentiation into the wireless networks and with it the *proportional QoS model*. The proportional service differentiation was introduced by Dovrolis et al [7] as a per-hop behavior for DiffServ (Differentiated Services) wired networks. It specifies the proportional QoS performance metrics and proportional relation between the QoS performance metrics. For example, if we consider the delay QoS metric, then the delay differentiation can be defined as follows. Assuming delay differentiation in a *class-based network* with ***C*** service classes, then the differentiation model imposes the following proportional constraints for all pairs of classes:

$$\frac{\bar{d}_i(t,t+\tau)}{\bar{d}_j(t,t+\tau)} = \frac{\delta_j}{\delta_i}, \text{ for all } i \neq j \quad \text{and} \quad i,j \in \{1,2,...,C\}, \qquad (1.1)$$

where δ_i is the service differentiation parameter for class i, and $\bar{d}_i(t,t+\tau)$ is the *average* delay for class i $(i = 1,2,\ldots, C)$, in time interval $(t, t + \tau)$, where τ is the monitoring time scale [2].

The major idea behind the proportional service differentiation is that even if the resources change over time, the qualities of flows with respect to each other stay constant, i.e., the quality ratio between classes remains constant in various time-scales. For example, if we consider two audio flows f_1 and f_2 with their average end-to-end delays between mobile hosts n_1 and n_2 and AP (with wireless links (n_1, AP) and (n_2, AP)), and d_1 and d_2 being in 2:1 proportion, then if these flows started with average delays $d_1 = 160$ ms and $d_2 = 80$ ms within the first time interval (0, τ), and in the next time interval (τ, 2τ), resource contention increases, i.e., resource availability changes, then d_1 and d_2 will change in 2:1 proportion to average delays $d_1 = 200$ ms and $d_2 = 100$ ms.

We discuss the proportional service differentiation model and their QoS parameter representation in the wireless LAN in more detail in Chapter 4. Note that we consider the concept of flows with proportional QoS not only within one node, but also for flows among different pairs of nodes. While proportional delay has been studied in wired networks extensively, there are still challenges in wireless LANs to provide these types of guarantees.

1.3.3 Statistical QoS

Statistical QoS model is the most used and most representative model in wireless networks, where service providers aim to satisfy and conform to average QoS requirements or to desired QoS requirements with a certain probability. It means that in case of statistical QoS representation, we want to express the probability that QoS requirement is satisfied. For example, probability P = Prob ($d \in [d_{min}, d_{max}]$) means the probability P that delay QoS $d \in [d_{min}, d_{max}]$ requirement is satisfied. Note that multimedia applications can tolerate some delay violations with probability $1 - P$, where packets' delays fall out of the range $[d_{min}, d_{max}]$, but it needs to be controlled. Hence, this type of representation is especially useful for multimedia applications with soft-real-time guarantees requirements, where adaptive behavior is acceptable, and QoS values can be satisfied with high probability. For example, multimedia playback application can choose delay requirement d_{req} for multimedia flow's packets with

$$d_{req} = \min(d \in D \mid D = \{d_1, d_2, ..., d_N\} \text{ and } \mathrm{Prob}(d \in [d_{min}, d_{max}]) > 95\%), \qquad (1.2)$$

where D is the set of multiple delay levels, given by the application. Often, QoS-aware resource functions, services, and protocols work with statistical average values and estimates of QoS parameters. For example, if a QoS operation measures N round-trip delay measurements $(d_1, d_2, \ldots, d_N)$, then it can use $d_{avg} = \frac{1}{2N}\sum_{i=1}^{N} d_i$, as the average value to estimate *end-to-end delay* [8]. Another example can be the *loss rate estimate* for flow f: $L(f) = 1 - j/k$, where k is the number of packets a flow attempts to transmit in a time interval T, and j is the number of attempts that result in successful packet transmission, while $k - j$ packets are dropped by the MAC protocol [9].

1.4 FUNCTIONAL COMPONENTS TOWARD QOS PROVISIONING

As we discussed, above, due to limited wireless resources and other challenging characteristics of wireless networks, to achieve QoS is hard. If the service provider aims to deliver and achieve statistical or proportional QoS requirements, given by a user, it is necessary to include certain *fundamental functions* within the resource management and protocols of wireless networks to provide these QoS guarantees. We will discuss four major necessary functional components: (1) *admission control,* (2) *QoS enforcement* via scheduling, bandwidth management, rate control, and incentives, (3) *QoS adaptation* via estimation and feedback control, and (4) *QoS-aware routing*. Depending on the level and strictness of the QoS requirements, the wireless service providers might provide all four QoS-aware functions in the protocol stack or just three QoS-aware functions (scheduling, adaptation, and routing) or just two QoS-aware functions (scheduling and adaptation) or just one QoS-aware function (adaptation). Using all four QoS-aware functions, we get a very *pro-active approach* and get closest with the network performance to the desired QoS requirements. However, this is also the most expensive approach. Using three, two, or one QoS-aware functions, we get *reactive approaches*, since we do not perform any admission control at the entrance of a QoS-sensitive flow and do not keep any state per flow. What we do is react to the resources availability and adapt each packet of a flow to the new resource situation. In this case, we must monitor the network performance of all flows, and based on their performance over given resources availability, we then adapt/determine the needed priorities and QoS level for scheduling, choice of a service class, and routing, respectively. The less QoS-aware functions we deploy in the wireless protocol stack, the less costly the wireless solutions will be, but the farther we get from the conformance of user desired QoS requirements.

The first QoS-aware functional component is *admission control* that admits each delay/bandwidth (BW)-sensitive flow at the entry into the wireless system. This component keeps track of how many flows have been admitted, how much resource was committed to admitted delay/BW-sensitive flows, and how much resource is still available for new flows. If there are not enough resources for new incoming delay/BW-sensitive flows, these new delay/BW-sensitive flows must be rejected to maintain guarantees made to already admitted flows.

The second QoS-aware functional component is *QoS enforcement*, which allocates resources to admitted time-sensitive flows and enforces that admitted flows use only agreed up amount of resources, and best effort flows, which do not go through admission control do not degrade the performance of admitted flows. It means that this component, for example, regulates sending rates for time-sensitive flows, as well as for the best effort flows to prevent degradation of QoS for admitted flows. One important design aspect for QoS enforcement will come out in this series of chapters and that is the *cross-layer design* to enforce end-to-end QoS requirements. The cross-layer design has been shown over and over again as one of the most important design choices toward efficient wireless networks and toward delivery of time-sensitive traffic with proportional and statistical QoS guarantees [88].

The third component is *QoS adaptation*, which deals with QoS violation, QoS and resource adjustments, and conflict resolution. This is a very important component in delivery of QoS in wireless networks, since the expectation is that this functional component carefully monitors resource usage and QoS levels and reacts with *adaptive policies* in case of QoS violations and changes in resource availability for time-sensitive flows. This means adjustment of QoS levels, re-allocation of resources, adjustment of priorities are part of the QoS adaptation design. Note that similarly to QoS enforcement, *cross-layer design* is a very important dimension in achieving effective QoS adaptation in wireless networks.

The fourth component is *QoS-aware routing*, which is used in multi-hop wireless networks and aims to find the best route according to the QoS requirements of time-sensitive flows through which then data will be disseminated. QoS-aware routing is very hard especially in MANETs. The reason is that in MANETs, one cannot establish a "reserved QoS route" as it is done in wired networks, and one cannot keep any routing states and the corresponding QoS levels reserved in each node, since nodes move and information become obsolete very quickly. What one can do toward QoS requirements is either to make the best possible QoS-aware routing decision locally at each node based on neighboring nodes routing/QoS status information or one assumes certain predictable/regular mobility patterns of mobile nodes, and then having location context available, one can provide location-guided QoS-aware routing.

In the next series of chapters, we will show selectively case studies of mathematical frameworks, algorithms, and protocol designs for these functional components in different types of wireless networks (as characterized in Section 1.2) and discuss these approaches in terms of their potentials and limitations when coping with QoS challenges. In Chapter 2, we will present the *utility-based approaches* toward admission control as well as *price-based resource allocation aspects*. In Chapter 3, we will discuss *rate and bandwidth management* algorithms such as rate control and bandwidth estimation to achieve statistical and proportional QoS guarantees. Chapter 4 will outline *delay management* techniques, concentrating on delay QoS adaptation approaches that take into account monitoring and new estimations of resources to adjust to delay QoS violations, feedback control to minimize the level of delay QoS degradation, cross-layer scheduling and priority mapping, and cross-layer design for QoS adaptation across application, middleware, network layers as well as across application, OS, and network layers. Chapter 5 describes different dimensions of QoS routing ranging from predictive location-based source routing algorithm, fault-tolerant routing protocol design in adversarial and overload environments, to energy-efficient utility-based routing optimization framework.

It is important to stress that throughout these chapters, different layers of the wireless network protocol stack will be affected, and we present cross-layer algorithms as well as individual approaches that can reside within MAC, network, transport, middleware/session, and application layers to delivery wireless network performance toward users' desired QoS requirements. Note that this lecture is not aimed to be an exhaustive survey of all QoS techniques that the wireless research community proposed and investigated over the last 10–15 years. Rather, our aim is to show on few solution examples the depth of the QoS problem difficulties and possible directions that one may take to search for QoS solutions. We include solutions such as QoS-aware routing functions, cross-layering approaches, prioritization techniques, location-aware protocols, location-delay predictive schemes, and interference awareness toward delivery of QoS requirements over contention-based wireless networks residing within unlicensed spectrum.

• • • •

CHAPTER 2

QoS-Aware Resource Allocation

2.1 INTRODUCTION

In this chapter, we explore the Quality of Service (QoS)-aware resource allocation in multi-hop wireless ad hoc networks. The reason that we investigate the QoS-aware resource allocation problem in multi-hop wireless ad hoc networks is that these are complex networks within which we can manifest extensively the difficulties, inherent features, and concepts of the QoS-aware resource allocation problem and solution spaces. We present the *utility-based and price-based problem formulations* of the optimal resource allocation with QoS constraints such as *rate*. From these problem formulations, one can then derive and/or modify problem formulations and QoS constraints to achieve appropriate resource allocation conforming to QoS requirements in variety of wireless network topologies as we will discuss in Chapters 3, 4, and 5.

We will consider a wireless ad hoc network where each node forwards packets to its peer nodes and each *end-to-end flow* traverses multiple hops of wireless links from a source to a destination. The unique characteristics of multi-hop wireless networks, compared to wireline networks, show that flows compete for shared channels if they are within the interference ranges of each other (*contention in the spatial domain*). This presents the problem of designing a topology-aware resource allocation algorithm that is both optimal with respect to *resource utilization and its conformation to QoS requirements (constraints)* and *fair* across contending multi-hop flows.

The discussion of the problem formulation for the optimal resource allocation will include first the description of the system model, needed concepts, and definitions in Section 2.2, and then, Section 2.3 presents a discussion of a utility-based description of the problem and its dual price-based description of the problem. In Section 2.4, we will discuss some practical considerations related to the resource allocation problems, especially the difference in resource allocation between wireline and wireless networks and the impact of interference ranges on the resource allocation problem in wireless networks. We will conclude the chapter in Section 2.5. The chapter will use notations and definitions shown in Table 2.1. Some basic concepts and their explanations in this chapter are shown in Table 2.2 (at the end of Section 2.2).

TABLE 2.1: Notations for Chapter 2.

NOTATIONS	DEFINITIONS
$G = (V,E)$ with $E \subseteq 2^V$, $e = \{i,j\}$, $i, j \in V$	Static multi-hope wireless ad hoc network represented as graph G with vertices V and edges E, edge $e \in E$ being between nodes i and j
d_{tx}	Wireless transmission range of node $i \in V$
d_{int}	Wireless interference range of node $i \in V$
d_{ij}	Distance between two nodes $i, j \in V$
SNR_{ij}	Signal-Noise-Ratio at node j with respect to transmission from node i to j
$\text{SNR}_{\text{thresh}}$	Minimum threshold of the Signal-Noise-Ratio in the wireless network under which transmission happens
$f \in F$	End-to-end flow f as part of the set of end-to-end flows F. Each f goes through multiple hops, passing a set of wireless links $E(f)$.
$S \subseteq E$, $s \in S$	Subset of wireless links in G such that each of the wireless links in S carries at least one subflow s.
$G_c = (V_c, E_c)$	Wireless link contention graph
$\boldsymbol{y} = (y_s, s \in S)$	Rate vector, where y_s is the rate on link s
C	Channel capacity
$q \in Q$	Maximal clique q as part of the set of all maximal cliques Q
$\text{V}(q) \subseteq S$	For clique q, the set of vertices is $V(q)$
$\boldsymbol{R} = \{R_{qf}\}$	Clique-flow matrix that describes the resource constraints on rate allocation among flows
$R_{qf} = \|\text{V}(q) \cap E(f)\|$	Number of subflows that a flow f has in the clique q.
$x = (x_f, f \in F)$	Feasible rate allocation for all flows f
$U_f(x_f) : \Re^+ \rightarrow \Re^+$	Utility function for end-to-end flow f presents degree of satisfaction of its corresponding end user

TABLE 2.1: (*continued*)

NOTATIONS	DEFINITIONS
$x_f \in I_f$, $I_f = [m_f, M_f]$	Feasible rate for one flow f, where x_f values are taken from the interval I_f with lower bound m_f (minimum rate bound) and upper bound M_f (maximum rate bound)
μ_q	Lagrange multiplier for clique q
$\mu_q = (\mu_q, q \in Q)$	Vector of Lagrange multipliers
C_q	Clique capacity

2.2 SYSTEM MODEL

We consider a wireless ad hoc network $G = (V, E)$ that consists of a set of nodes V as shown in Figure 2.1a. The network topology shows seven nodes $V = \{1,2,3,4,5,6,7\}$, bidirectional edges $E = \{(1,2), (2,1), (2,3), (3,2), (3,4), (4,3), (4,5), (5,4), (3,6), (6,3), (6,7), (7,6)\}$. Each node $i \in V$ has a *transmission range* d_{tx}, and an *interference range* d_{int}, which can be larger than d_{tx}. In Figure 2.1, we assume that the transmission range is the same as the interference range. A wireless node $i \in V$ may establish an end-to-end flow f_i, with a certain rate x_i (*QoS requirement*) to another node. Flow f_i is assumed to be *elastic*; it means it requires a minimum rate x_i^m and a maximum rate x_i^M, i.e., $x_i^m \leq x_i \leq x_i^M$. In general, flow f_i flows through multiple hops in the network, passing a set of wireless links. A single-hop data transmission along a particular wireless link is referred to as a subflow s and is part of a flow. Several subflows from different flows along the same wireless link form an aggregated subflow. In Figure 2.1a, we have seven end-to-end flows $\{f_1, f_2, f_3, f_4, f_5, f_6, f_7\}$, where end-to-end

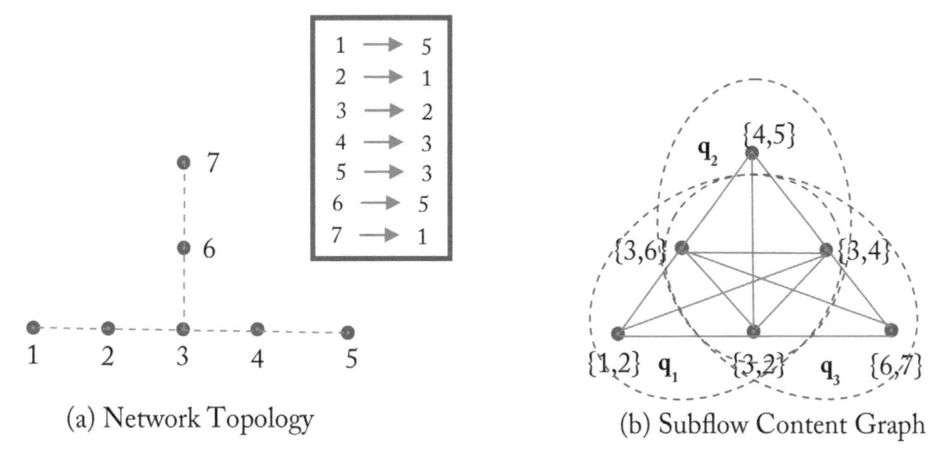

FIGURE 2.1: Example of wireless ad hoc network and its subflow contention graph [90].

flow f_1 goes from nodes 1 to 5, flow f_2 from nodes 2 to 1, flow f_3 from nodes 3 to 2, flow f_4 from nodes 4 to 3, flow f_5 from nodes 5 to 3, flow f_6 from nodes 6 to 5, and flow f_7 from nodes 7 to 1.

In such a network, nodes compete for two types of resources: *shared channel* and *individual nodes's relaying cost* (e.g., energy). The availability of these resources constrains the solution space of resource allocations. In this chapter, we will concentrate on formulating the resource allocation problem for *shared channel* for clarity presentation reasons. The full problem formulation for both resources (shared channel and energy) can be found in [90].

2.2.1 Shared Wireless Channel

The shared-medium multi-hop nature of wireless ad hoc networks presents unique characteristics of location-dependent contention and spatial reuse of spectrum. It means, packet transmission in such a network is subject to *location-dependent contention*. Compared with wireline networks where flows contend only at the routers with other simultaneous flows through the same router (contention in the time domain), the unique characteristics of multi-hop wireless networks show that flows also compete for shared channel bandwidth if they are within the transmission range d_{tx} of each other (contention in the spatial domain).

In particular, two subflows contend with each other if either the source or destination of one subflow is within the transmission range d_{tx} of the source or destination of the other. (Note that if we assume that $d_{\text{int}} > d_{\text{tx}}$, then contention model can be straightforwardly extended.) The locality of wireless transmissions implies that the degree of contention for the shared medium is location-dependent. On the other hand, two subflows that are geographically far away have the potential to transmit simultaneously, reusing the wireless channel. The contention model can be represented as a contention graph. Figure 2.1b shows a subflow contention graph under the assumption of flows transmitted over the network topology in Figure 2.1a. The contention graph in Figure 2.1b will have an edge if two subflows contend. For example, a subflow s_1 from nodes 1 to 2, $s_1 = (1,2)$, in network topology Figure 2.1a contends with a subflow s_2 that goes from nodes 3 to 2, $s_2 = (3,2)$. Hence, the contention graph in Figure 2.1b will have vertices s_1 and s_2 that represent edges in G (i.e., (1,2) and (3,2) in our example) and an *edge in contention graph* represents contention relation between the two considered edges in G (i.e., there will be in edge in the contention graph G_c between s_1 and s_2 since there is contention between subflows s_1 and s_2 going over edges (1,2) an (3,2) in G).

The location-dependent contention for packet transmission manifests itself in wireless networks at the protocol level and at the physical level [10], generally referred to as the *protocol model* and the *physical model*. In the case of a single wireless channel, these two models have the following characteristics:

a. *Protocol Model*: The tranmission from node i to j, $(i, j \in V)$ is successful if (1) the distance d between these two nodes d_{ij} satisfies $d_{ij} < d_{tx}$; (2) any node $k \in V$, which is within the interference range of the receiving node j, $d_{kj} \leq d_{\text{int}}$ is not transmitting. This model can be further refined toward the case of IEEE 802.11 MAC protocols, where the sending node i is also required to be free of interference as it needs to receive the link layer acknowledgment from the receiving node j. Specifically, any node $k \in V$, which is within the interference range of the nodes i or j (i.e., $d_{ki} \leq d_{\text{int}}$ or $d_{kj} \leq d_{\text{int}}$), is not transmitting.

b. *Physical Model*: This model is directly related to the physical layer characteristics. The transmission from node i to j is successful if the signal-to-noise ratio at the node j, SNR_{ij}, is not smaller than a minimum threshold: $SNR_{ij} \geq SNR_{\text{thresh}}$.

We focus our attention on addressing the problems of QoS-aware resource allocation based on the *protocol model* with particular interest in IEEE 802.11-style MAC protocols due to their popular deployment in realistic wireless systems.

2.2.2 Contention Model and Resource Constraints

In this section, we present the contention model and resource constraints discussed above more formally. Under the protocol model, a wireless ad hoc network will be modeled as a bidirectional graph $G=(V,E)$. $E \subseteq 2^V$ denotes the set of wireless links, which are formed by nodes from V that are within the transmission range of each other. A wireless link $e \in E$ is represented by its end nodes i and j, i.e., $e = \{i,j\}$ as we discussed above and showed an example in Figure 2.1a. The set of end-to-end flows will be denoted as F. Each flow $f \in F$ goes through multiple hops in the network, passing a set of wireless links $E(f)$. A single hop data transmission in the flow f along a particular wireless link is referred to as a subflow s of f. Obviously, there may exist multiple subflows along the same wireless link. We denote $S \subseteq E$ as a set of wireless links in G, such that each of the wireless links in S carries at least one subflow s, i.e., the wireless link is not idle.

Flows in a wireless ad hoc network *contend* for shared resources in a location-depedent manner: two subflows contend with each other if either the source or destination of one subflow is within the interference range (d_{int}) of the source or destination of the other. Among a set of mutually contending subflows, only one of them may transmit at any given time. Thus the aggregated rate of all subflows in such a set may not exceed the channel capacity. Formally, we consider a *wireless link contention graph* [11] $G_c = (V_c,E_c)$, in which vertices correspond to the wireless links (i.e., $V_c = S$), and there exists an undirected edge between two vertices if the subflows along these two wireless links contend with each other.

For example, in Figure 2.1b, we have $V_c = \{\{1,2\}, \{3,2\}, \{6,7\}, \{3,6\}, \{3,4\}, \{4,5\}\}$ and $E_c = \{(\{1,2\}, \{3,2\}), (\{3,2\}, \{6,7\}), (\{3,2\}, \{3,4\}), (\{3,2\}, \{3,6\}), (\{1,2\}, \{3,6\}), (\{1,2\}, \{3,4\}), (\{3,6\}, \{6,7\}), (\{3,6\}, \{3,4\}), (\{3,6\}, \{4,5\}), (\{4,5\}, \{3,2\}), (\{4,5\}, \{3,4\}), (\{3,4\}, \{6,7\})\}$.

In an undirected graph, a complete subgraph is called a *clique*, i.e., a subset of its vertices form a clique if every two vertices in the subset are connected by an edge. Figure 2.2 shows a set of cliques.

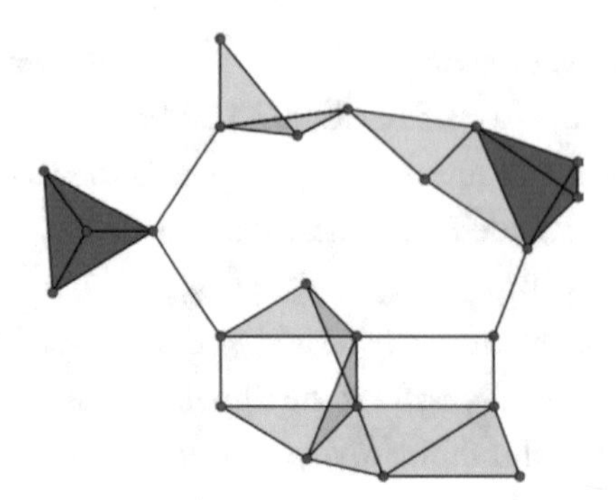

FIGURE 2.2: Examples of cliques in an undirected graph [91]. Light blue areas represent the three-vertex cliques, dark blue areas represent maximal four-vertex cliques.There are also two-vertex cliques (its edges) and one-vertex cliques (its vertices).

In Figure 2.1b, we have many cliques as well. For example, the subset of vertices {(1,2), (3,2), (3,6)} creates a clique since all vertices in the subset are connected. A special clique is the *maximal clique*. A *maximal clique* is defined as a clique that is not contained in any other cliques, i.e., it is a clique that cannot be extended by including one more adjacent vertex. In Figure 2.1b, we have three maximal four-vertex cliques q_1, q_2, q_3. $q_1 = \{\{1,2\}, \{3,2\}, \{3,4\}, \{3,6\}\}$, $q_2 = \{\{3,2\}, \{3,4\}, \{4,5\}, \{3,6\}\}$, $q_3 = \{\{3,2\}, \{3,4\}, \{3,6\}, \{6,7\}\}$. In a wireless link contention graph, the vertices in a maximal clique represent a maximal set of mutually contending wireless links, along which at most one subflow may transmit at any given time.

We proceed to consider the problem of allocation *rates* to wireless links (Note: The allocated rate is a QoS requirement that specifies the constraint on bandwidth/capacity resource on the link, and we consider average rates for links since QoS for wireless networks is of statistical nature). Let us denote y_{ij} as the *aggregate rate* of all subflows along the *node* ***i*** and *node* ***j***. For example, y_{12} will be an aggregate rate of all subflows along nodes 1 and 2. From Figure 2.1a, we can see that $y_{12} = x_1 + x_2 + x_7$ because one subflow of f_1 (from nodes 1 to 5) with rate x_1 will go along nodes 1 and 2, subflow of flow f_2 (from nodes 2 to 1) with rate x_2 will go along nodes 1 and 2, and subflow of flow f_7 (from nodes 7 to 1) with rate x_7 will go along nodes 1 and 2. Other examples of aggregate rates would be $y_{23} = x_1 + x_3 + x_7$ along nodes 2 and 3 or $y_{36} = x_6 + x_7$ along nodes 3 and 6.

We claim that an aggregate rate allocation $\boldsymbol{y} = (y_s, s \in S)$ is *feasible*, if there exists a *collision-free* transmission schedule that allocates y_s to s. Formally, if a rate allocation $\boldsymbol{y} = (y_s, s \in S)$ is feasible, then the following condition is satisfied [12]:

$$\forall q \in Q, \sum_{s \in V(q)} y_s \leq C; \tag{2.1}$$

where Q is the set of all maximal cliques in G_c, and C is the channel capacity. For clique q in the wireless link contention graph G_c, $V(q) \subseteq S$ is the set of its vertices.

For example, $V(q_1)$, $V(q_2)$, $V(q_3)$ from Figure 2.1b are as follows: $V(q_1) = \{\{1,2\}, \{3,2\}, \{3,6\}, \{3,4\}\}$, $V(q_2) = \{\{3,2\}, \{3,6\}, \{3,4\}, \{4,5\}\}$, $V(q_3) = \{\{6,7\}, \{3,2\}, \{3,6\}, \{3,4\}\}$.

Equation 2.1 gives an upper bound on the rate allocations to the wireless links. In practice, however, such a bound may not be tight, especially with carrier-sensing multiple-access-based wireless networks as IEEE 802.11 networks. In this case, we introduce C_q, the *achievable capacity* at a clique q. More formally, if $\sum_{s \in V(q)} y_s \leq C_q$, then $\boldsymbol{y} = (y_s, s \in S)$ is feasible. One can see that each maximal clique may be regarded as an *independent channel* resource unit with capacity C_q. Hence, we will use a *maximal clique* as a *basic resource unity* for defining the optimal resource allocation problem, using utility-based problem description as well as price-based problem description.

As an example, we will consider Figure 2.1b to show the achievable capacity at each clique q_1, q_2, q_3, where in each clique q_1, q_2, q_3, the aggregate rate cannot exceed the channel capacity C_1, C_2, C_3, i.e.,

$$y_{12} + y_{32} + y_{34} + y_{36} \leq C_1$$

$$y_{32} + y_{34} + y_{45} + y_{36} \leq C_2$$

$$y_{32} + y_{34} + y_{36} + y_{67} \leq C_3.$$

To consider description of *resource constraints* on rate allocation among flows, we define a *clique-flow matrix* $\boldsymbol{R} = \{R_{qf}\}$, where $R_{qf} = |V(q) \cap E(f)|$ represents the number of subflows that a flow f has in the clique q. If we treat a maximal clique as an independent resource, then the clique-flow matrix $\boldsymbol{R}$ represents the "resource usage pattern" of each flow. Let the vector $\mathbf{C} = (C_q, q \in Q)$ be the vector of achievable channel capacities in each of the cliques. In a wireless ad hoc network $G = (V,E)$ with set of flows F, there exists a feasible rate allocation $\boldsymbol{x} = (x_f, f \in F)$, if and only if $Rx \leq C$. This observation gives the constraints with respect to rate allocation to end-to-end flows in wireless ad hoc networks.

As an example, let us consider the end-to-end flow rate allocation from Figure 2.1, where the resource constraint $Rx \leq C$ imposed by the shared wireless channel is as follows: $\begin{pmatrix} 3 & 1 & 1 & 1 & 1 & 2 & 3 \\ 3 & 0 & 1 & 1 & 2 & 3 & 2 \\ 2 & 0 & 1 & 1 & 1 & 2 & 3 \end{pmatrix} \cdot x \leq C$, where the matrix represents the clique-flow matrix $\boldsymbol{R}$ for our example in Figure 2.1.

2.2.3 Integrating Example

We present another example in Figure 2.2 [13] to illustrate all the concepts together as discussed in Sections 2.2.1. and 2.2.2. Furthermore, in Table 2.2, we will summarize all concepts in a tutorial manner.

Figure 2.3a shows the topology of the wireless ad hoc network as well as the ongoing flows f_1, f_2, f_3, and f_4 (e.g., flow f_4 goes from node 5 to node 4). The corresponding wireless link contention graph for the Figure 2.3a flows and topology is shown in Figure 2.3b. In Figure 2.3b, we assume that the interference range d_{int} is the same as the transmission range d_{tx} ($d_{int} = d_{tx}$). In Figure 2.3c, we show the link contention graph when the interference range is twice as large as the transmission range ($d_{int} = 2^*d_{tx}$).

In this example, there are four end-to-end flows f_1 = {{1,2}, {2,3}, {3,4}, {4,5}}, f_2 = {{7,6}, {6,3}}, f_3 = {{6,3}, {3,2}, {2,1}}, and f_4 = {{5,4}}. In Figure 2.3b, there are three maximal cliques in the contention graph: q_1 = {{1,2}, {3,2}, {3,4}, {3,6}}, q_2 = {{3,2}, {3,4}, {4,5}, {3,6}}, and q_3 = {{3,2}, {3,4}, {3,6}, {6,7}}. In Figure 2.3c, there is only one maximal clique q_1 = {{1,2}, {3,2}, {3,4}, {3,6}, {4,5}, {6,7}} because $d_{int} = 2^*d_{tx}$.

We use y_{ij} to denote the aggregated rate of all subflows along wireless link $\{i,j\}$. For example, $y_{12} = x_1 + x_3$, $y_{36} = x_2 + x_3$ with x_1 being the individual rate of flow f_1, x_2 being the rate of flow f_2 and x_3 being the rate of flow f_3. In each clique, the aggregated rate may not exceed the corresponding channel capacity. In particular, when $d_{int} = d_{tx}$, then

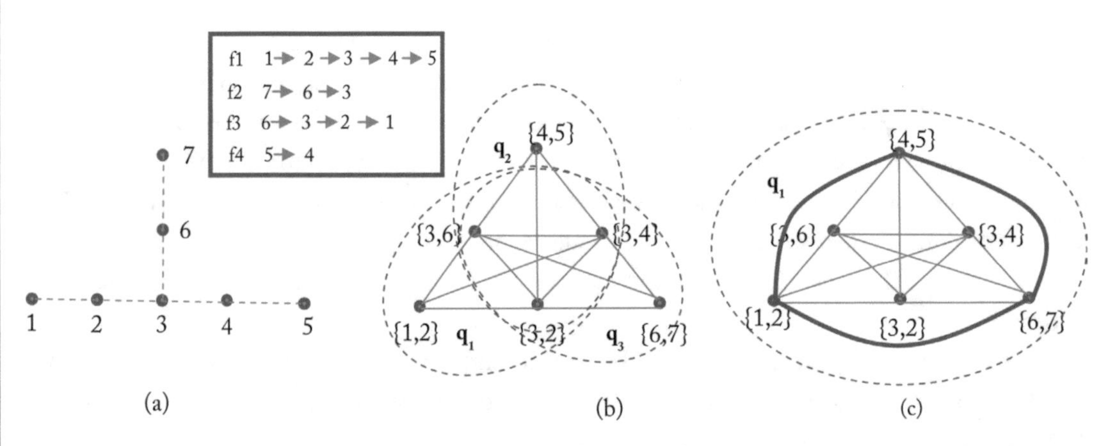

FIGURE 2.3: Wireless Link Contention Graph and Resource Constraints: Example with four flows f_1, f_2, f_3, f_4 shown in Figure 2.2a network topology of wireless ad hoc nodes; and corresponding cliques for the four flows in cases Figure 2.2b $d_{int} = d_{tx}$ and Figure 2.2c $d_{int} = 2^*d_{tx}$ [13].

$$y_{12} + y_{32} + y_{34} + y_{36} \leq C_1;$$

$$y_{32} + y_{34} + y_{45} + y_{36} \leq C_2;$$

$$y_{32} + y_{34} + y_{36} + y_{67} \leq C_3;$$

when $d_{\text{int}} = 2^*d_{\text{tx}}$, then $y_{12} + y_{32} + y_{34} + y_{45} + y_{67} \leq C_1'$.

When it comes to end-to-end flow rate allocation, the resource constraint imposed by shared wireless channels is as follows. When $d_{\text{int}} = d_{\text{tx}}$,

$$\begin{pmatrix} 3 & 1 & 3 & 0 \\ 3 & 1 & 2 & 1 \\ 2 & 2 & 2 & 0 \end{pmatrix} x \leq C$$

and when $d_{\text{int}} = 2^*d_{\text{tx}}$, then $(4 \;\; 2 \;\; 3 \;\; 1) \leq C'$.

As we can see, the unique characteristics of location-dependent contention in wireless ad hoc networks imply a fundamentally *different resource model* compared to the case of wireline networks. In wireline networks, the capacity of a link represents the constraints on flows contending for its bandwidth which is independent from other links. However, in case of wireless ad hoc networks, the capacity of a wireless link is interrelated with other wireless links in its neighborhood. This strong interference and dependency must be taken into account with respect to models of resource constraints and allocations in wireless networks as we show in Section 2.3, discussing utility-based description and price-based description of optimal resource allocation. In Table 2.2, we summarize concepts and the corresponding definitions used in this chapter.

TABLE 2.2: Glossary of concepts.

CONCEPT	DEFINITION
Clique	In an undirected graph, a complete subgraph is called a *clique*, i.e., a subset of its vertices form a clique if every two vertices in the subset are connected by an edge
Maximal clique	A *maximal clique* is defined as a clique that is not contained in any other cliques, i.e., it is a clique that cannot be extended by including one more adjacent vertex

(*continued*)

TABLE 2.2: (*continued*)

CONCEPT	DEFINITION
Wireless link contention graph	*Wireless link contention graph* $G_c = (V_c, E_c)$, is a graph in which vertices correspond to the wireless links (i.e., $V_c = S$), and there exists an undirected edge between two vertices if the subflows along these two wireless links contend with each other
Utility function	Measure of relative satisfaction; social welfare
Pareto optimality	Measure of efficiency; Pareto optimal outcome is one outcome where no one could be made better off without making someone else worse off
Proportional fairness	A vector of rates $\mathbf{x} = (x_f, f \in F)$ is *proportionally fair* if it is feasible (that is $x \geq 0$ and $Rx \leq C$) and if for any feasible vector $\mathbf{x}^*$, the aggregate of proportional changes is zero or negative: $\sum_{f \in F} \frac{x_f^* - x_f}{x_f} \leq 0$
Max-min fairness	A vector of rates $\mathbf{x} = (x_f, f \in F)$ is *max-min fair* if it is feasible (that is $x \geq 0$ and $Rx \leq C$) and if for each $f \in F\mathbf{x}$ cannot be increased (while maintaining feasibility) without decreasing x_f^* for some f^* for which $x_f^* \leq x_f$ [93]
Lagrange multiplier	Method of Lagrange multipliers provides a strategy for finding the maxima and minima of a function subject to constraints. For example, let us consider that the optimization problem maximizes $f(x,y)$, subject to $g(x,y) = c$. We introduce a new variable (μ), called Lagrange multiplier and study the Lagrange function defined by $\Lambda(x, y, \mu) = f(x, y) + \mu \cdot (g(x, y) - c)$, where the μ term may be either added or substracted. Note that the method of Lagrange multipliers yields a necessary condition for optimality in constrainted problems, since not all points (x,y,μ) yield a solution to the original problem. Figure 2.4 shows visually the optimization problem

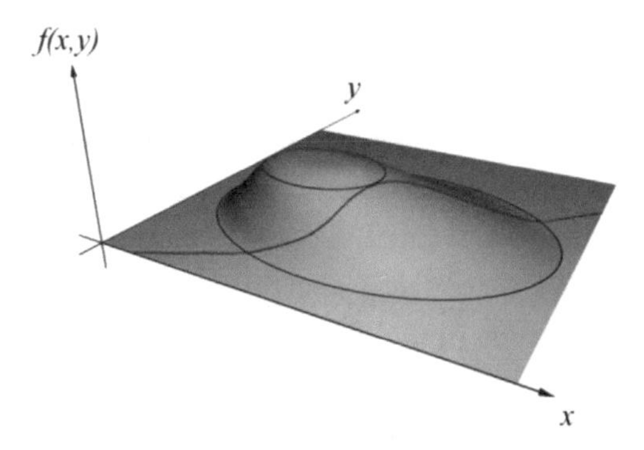

FIGURE 2.4: Visualization of optimization problem: Goal is to find x and y that will maximize $f(x,y)$ subject to constraint $g(x,y) = c$ (shown in red) [92].

2.3 RESOURCE ALLOCATION PROBLEM FORMULATIONS

To address the QoS-aware resource allocation problem for multi-hop wireless ad hoc networks, we will first motive the problem. After the intuitive motivation of the problem, we will present a more formal formulation and discuss the utility-based resource allocation problem and its dual formulation, the price-based resource allocation.

The unique characteristics of multihop wireless networks show that flows compete for shared channel if they are within the interference range of each other (contention in the spatial domain). This presents the problem of designing a topology-aware resource allocation algorithm that is both *optimal* with respect to resource utilitization and QoS provisioning, and *fair* across contending multihop flows. *Fair packet scheduling mechanisms* [94, 95, 96] have been shown to perform effectively in providing fair shares among single hop flows in wireless ad hoc networks and in balancing the tradeoffs between fairness and resource utilization. While these mechanisms are sufficient for maintaining basic fairness properties among localized flows, they do not coordinate intraflow (subflow) resource allocations between upstream and downstream hops of an end-to-end flow and thus will not be able to achieve *global optimum* with respect to resource utilization and fairness.

Owing to the complexities of such intraflow coordination, one appropriate strategy to solve the end-to-end flow resource allocation problem is to use *price-based strategy*, where prices are computed as signals to reflect relations between resource demands and supplies and are used to coordinate resource allocations at multiple hops. Previous research in *wireline network* pricing [e.g., 14, 15] has shown that pricing is effective as a means to arbitrate resource allocation. In these research results, a *shadow price* is allocated with a wireline link to reflect relations betwee the traffic load of the link and its bandwidth capacity. A *utility* is associated with an end-to-end flow to reflect its resource requirement. Transmission rates are chosen to respond to the aggregated price signals along end-to-end flows such that the net benefits (the difference between utility and cost) of flows are maximized. Moreover, by choosing appropriate utilities, various fairness models can be achieved.

Unfortunately, due to the fundamental differences between multihop wireless ad hoc networks and wireline networks, we cannot map directly existing *pricing theories* from wireline networks to multihop wireless networks. In multihop wireless networks, flows that traverse the same geographical vicinity contend for the same wireless channel capacity. This is in sharp contrast with wireline networks, where only flows that traverse the same link contend for its capacity. In wireline networks, we can associate shadow prices directly with individual links to reflect their resource demands and supply. This is not feasible in wireless networks with the presence of location-dependent contention.

In multi-hop wireless ad hoc networks, to answer the question "How much bandwidth should we allocate to each of the end-to-end flows so that scarce resources in a wireless network be optimally and fairly utilized?", one needs to modify the pricing framework and specifically tailor it to the contention model of wireless networks. This also means we need to establish *shadow prices* based on *maximal cliques in wireless link contention graphs*, rather than individual links, as we did in wireline networks. In such pricing framework, the price of an end-to-end multi-hop flow is to aggregate prices of all its subflows, while the price of each of the subflows is the sum of shadow prices of all maximal cliques that it belongs to.

With this pricing framework, by choosing the appropriate *utility functions*, the optimality of resource allocation in terms of rate QoS utilization and fairness will be then achieved by maximizing the aggregated utility across all flows. Next, we will present the utility-based formulation of the resource allocation problem.

2.3.1 Utility-Based Resource Allocation Problem

We associate each end-to-end flow $f \in F$ with a utility function $U_f : \Re^+ \rightarrow \Re^+$, which represents the degree of satisfaction of its corresponding end user when sending flow f and at its rate x_f. We make the following *assumptions* about the utility function U_f to make the problem of finding maximal aggregated utility $\sum_{f \in F} U_f(x_f)$ sound:

1. Over the interval $I_f = [m_f, M_f]$, the utility function U_f, with the elastic flow f's rate $x_f \in I_f$, is increasing, strictly concave, and twice continuously differentiable.
2. The curvature of U_f is bounded away from zero over I_f, i.e., $-U''_f(x_f) \geq 1/\kappa_f > 0$.
3. U_f is additive so that the aggregated utility of rate allocation $\boldsymbol{x} = (x_f, f \in F)$ is $\sum_{f \in F} U_f(x_f)$.

We investigate the problem of optimal rate allocation in the sense of *maximizing the aggregated utility function* $\sum_{f\in F} U_f(x_f)$, which is also referred to as the *social welfare* in the literature [13]. Such an optimization objective achieves *Pareto optimality* with respect to the resource utilization and also realizes different fairness models—including *proportional and max-min fairness* [14]—when appropriate utility functions are specified. The problem of optimal resource allocation in wireless ad hoc networks can be then formulated as a nonlinear optimization problem (primal problem):

$$\textbf{P: } \text{maximize} \sum_{f\in F} U_f(x_f) \tag{2.2}$$

$$\text{subject to } \mathbf{R}x \le \mathbf{C} \tag{2.3}$$

$$\text{over } x_f \in I_f \tag{2.4}$$

This optimization problem maximizes the aggregated utility function in Equation 2.2 with the resource capacity constraints shown in Equation 2.3 and coming from the shared wireless channel as discussed in Section 2.2. Note that we are dealing with *maximal cliques* and *clique-flow matrix* **R**, which represents the "*resource usage pattern*" of each flow as a constraint in (Equation 2.3). If we can optimize the problem **P**, we will get both optimal resource utilization and fair resource allocations among end-to-end flows spanning multiple hops. However, often arriving toward a solution of the problem **P** might be complex and difficult since finding the right utility functions is difficult, hence one might look at the dual problem **D** to problem **P** and consider pricing-based description **D** of the optimal resource allocation problem rather than utility-based description **P**.

2.3.2 Price-Based Resource Allocation Problem

To proceed toward the solution of the problem **P**, so that we achieve optimal resource allocation (and QoS with respect to rate allocation) in terms of both utilization and fairness, we will transform the problem **P** into its dual problem as follows. Owing to Assumption 1 in Section 2.3.1, the objective aggregated utility function $\sum_{f\in F} U_f(x_f)$ in Equation 2.2 is differentiable and strictly concave. Also, the feasible region of the optimization problem in Equations 2.3 and 2.4 is convex and compact. Hence, by nonlinear optimization theory, there exists a maximizing value of argument $\boldsymbol{x}$ for the above optimization problem. Let us consider the Lagrangian form of the optimization problem **P**:

$$L(x;\mu) = \sum_{f\in F} U_f(x_f) - x_f \sum_{q\in Q} \mu_q R_{qf} + \sum_{q\in Q} \mu_q C_q \tag{2.5}$$

where $\mu_q = (\mu_q, q \in Q)$ is a vector of Lagrange multipliers. Note that Equation 2.5 represents the method of Lagrange multipliers as we described in Table 2.2, where we have $f(x) = \sum_{f\in F} U_f(x_f)$,

$g(x) = \mathbf{R}x = \sum_{f\in F}\sum_{q\in Q} x_f R_{qf}$, $c = \mathbf{C} = \sum_{q\in Q} C_q$, and $\mu_q = (\mu_q, q \in Q)$ being the vector of Lagrange multipliers.

The Lagrange multiplier μ_q represents actually the cost of a unit flow *f*, accessing the wireless channel in the maximal clique *q*. μ_q is also called the *shadow price of the clique q* as we will explain below. For example, let us consider the Figure 2.5b with the flow *f* passing from node 1, through nodes 2, 3, 4, to node 5. The wireless network is assumed to have $d_{\text{int}} = d_{\text{tx}}$. The topology of the five wireless nodes, when transformed to the *wireless link contention graph* (as discussed in Section 2.2), will create two maximal cliques, $q_1 = \{\{1,2\}, \{2,3\}, \{3,4\}\}$ and $q_2 = \{\{2,3\}, \{3,4\}, \{4,5\}\}$ through which flow *f* passes. Each maximal clique q_1 and q_2 will have assigned their shadow prices μ_1 and μ_2. Note that Figure 2.5a shows the wireline network and here each subflow of the flow *f* (subflow from nodes 1 to 2, subflow from nodes 2 to 3, etc.) has its own shadow price μ_i (i.e., μ_1, μ_2, μ_3, μ_4). It is important to stress that in wireline networks, *links* get associated with the *cost (shadow price)* of a passing flow. In wireless network, *maximal cliques* get associated with the cost (shadow price) of a passing flow.

In the next step, we present a decentralized formulation of the resource allocation problem where knowledge of utility functions of all flows is not needed. The key to decentralization is to investigate its dual problem and to decompose the problem via *pricing*. Let us consider the dual problem **D** of the primal problem **P**:

$$\mathbf{D}: \min_{\mu\geq 0} D(\mu) \tag{2.6}$$

With

$$\begin{aligned} D(\mu) &= \max_{x_f\in I_f} L(x;\mu) \\ &= \sum_{f\in F} \max_{x_f\in I_f} \Big(U_f(x_f) - x_f \sum_{q:E(f)\cap V(q)\neq\varnothing} \mu_q R_{qf}\Big) + \sum_{q\in Q} \mu_q C_q \end{aligned} \tag{2.7}$$

Let us define

$$\lambda_f = \sum_{q:E(f)\cap V(q)\neq\varnothing} \mu_q R_{qf}. \tag{2.8}$$

From the dual problem **D**, the Lagrange multipliers μ_q can be interpreted as the implied cost of a unit flow, accessing the channel in the maximal clique *q*. It means, μ_q represents the *shadow price of the clique q* and λ_f, on the other hand, can be interpreted as the shadow price of the flow *f*. From Equation 2.8, we observe that flow *f* needs to pay for all the maximal cliques that it traverses. For each clique, the price to pay is the products of the number of wireless links that *f* traverses in this clique and the shadow price of the clique.

Alternatively, since

$$\lambda_f = \sum_{q:E(f)\cap V(q)\neq\varnothing} \mu_q R_{qf} = \sum_{s:s\in E(f)} \sum_{q:s\in V(q)} \mu_q \tag{2.9}$$

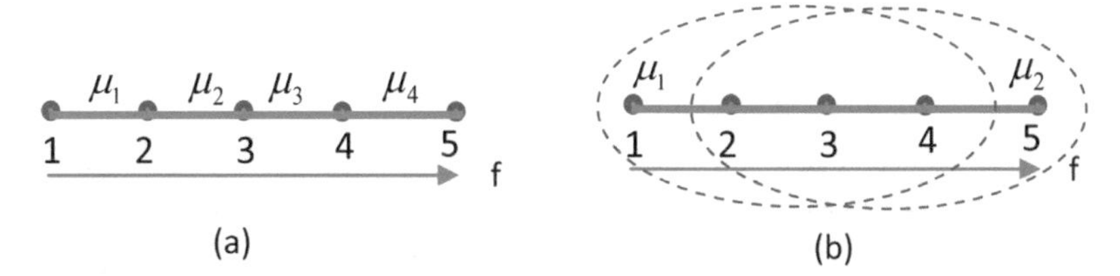

FIGURE 2.5: (a) Wireline network topology, (b) wireless network topology for flow *f* and its cost.

the shadow price of a flow is also the aggregated price of all its subflows *s*. For each subflow, its price is the aggregated price of all the maximal cliques that it belongs to.

We illustrate the price allocation on the example shown in Figure 2.5 [13] and compare the pricing of wireless and wireline networks. The wired network in Figure 2.5a is a chain topology consisting of four links, and their associated prices μ_1, μ_2, μ_3, μ_4. In this case, the price of the flow *f* is $\lambda_f = \sum_{l=1}^{4} \mu_l$.

Although the wireless network in Figure 2.5b has the same topology as the wired network, this topology will have the two maximal cliques q_1 and q_2, which represent its units for resource allocation. If the shadow prices of these two maximal cliques are μ_1 and μ_2, then the price of the flow *f* that traverses these two cliques is then given by $\lambda_f = 3\mu_1 + 3\mu_2$, which is the sum of the product of the number of subflows of *f* in each clique and the shadow price of this clique. The price can be also written as $\lambda_f = \mu_1 + (\mu_1 + \mu_2) + (\mu_1 + \mu_2) + \mu_2$, which is the sum of the prices of its subflows. The price of each subflow is the aggregated price of all the maximal cliques that it belongs to.

2.4 PRACTICAL ISSUES

We will briefly discuss and illustrate the meaning of the contention regions and the impact of interference in wireless ad hoc networks. We will consider the Figure 2.5 network topology and compare the equilibrium resource allocations of the four hop wireless ad hoc network with the four link wireline network. We will also compare the equilibrium resource allocations of two wireless ad hoc networks with different interference ranges.

First, if we compare the equilibrium resource allocations of two networks in Figure 2.5, the results show in Table 2.3 that the rate allocated to each flow in the ad hoc network is less than the rate allocated to the corresponding flow in wirelinenetworks. The difference lies in their different definitions of contention regions. In the wireline network, a wireline link represents a contention region, whose capacity is the link bandwidth. In the wireless ad hoc network, a wireless link is no longer a contention region. Instead, the set of wireless links, formally represented by a clique, constitutes the contention region and shares the channel capacity. Thus, with the same capacity of the

TABLE 2.3: Rate allocations (x_1, x_2, x_3, x_4, x_5) and equlibrium prices ($\mu_1, \mu_2, \mu_3, \mu_4$) in different networks specified in Figure 2.5

	x_1	x_2	x_3	x_4	x_5	μ_1	μ_2	μ_3	μ_4
Wireline network	0.4	1.6	1.6	1.6	1.6	0.625	0.625	0.625	0.625
Ad hoc network	0.133	0.8	0.4	0.4	0.8	1.25	1.25	N/A	N/A

wireless channel and the wireline link, the throughput of the wireless ad hoc network is lower than that of the wireline network.

Second, in the wireline network, the rates of all single-hop flows are the same. In the wireless ad hoc network, the rates of these flows are different. The reason is that in the wireline networks, flows f_2 through f_5 enjoy the same amount of resources, while in the wireless ad hoc network, due to location-dependent contention, f_3 suffers higher contention than f_2. This is also reflected through the prices that f_2 and f_3 need to pay. For f_2, the price is $\lambda_2 = \mu_1$, which equals to 1.25 at the equilibrium, while the price for f_3 is $\lambda_3 = \mu_1 + \mu_2$, which equals to 2.5.

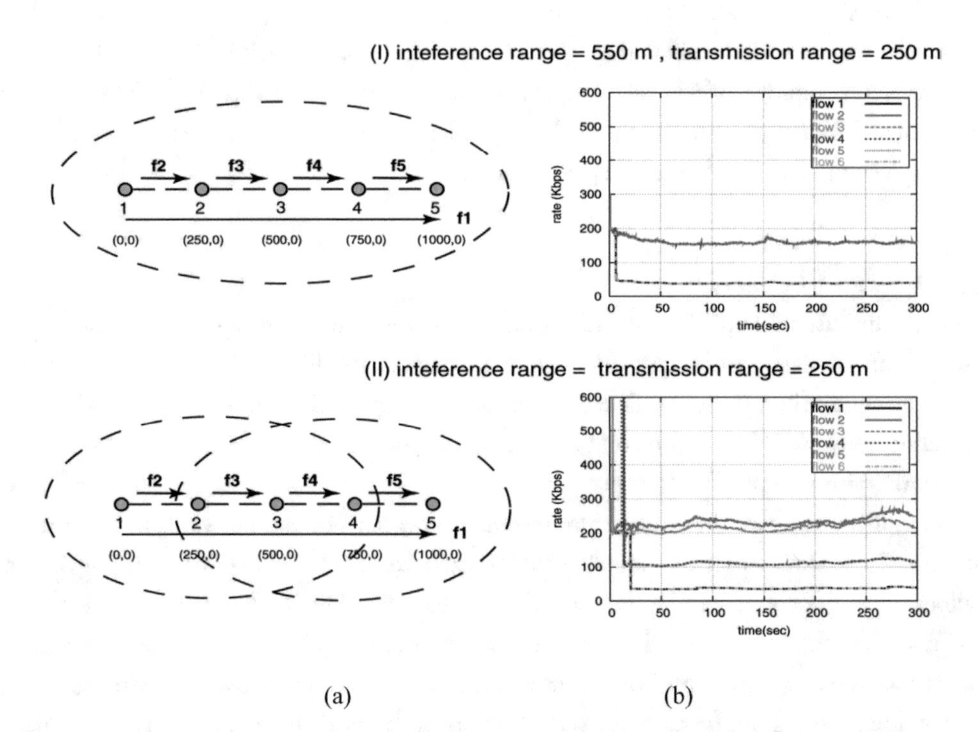

FIGURE 2.6: (a) Two wireless ad hoc networks with their cliques for different interference ranges. (b) Transmission rate for wireless flows f_2, f_3, f_4, f_5 under different interference ranges.

Third, in both networks, the equilibrium rate allocations for flows with different lengths are different. This is actually the result of *proportional fairness*. In particular, the longer the flow, the less the rate allocated. This observation is natural from the perspective of maximizing the aggregated utility. When the utility functions of all flows are the same, long flows consume more resources for an unit of utility increase. Hence, *short flows are favored.*

Fourth, to further illustrate the impact of interference in wireless ad hoc networks, we show and compare in Figure 2.6a and 2.6b two wireless ad hoc networks with different interference ranges. We observe that the resource allocations are different for the two networks. The reason behind this observation is that different interference ranges lead to different contention regions as shown in Figure 2.6a. When the interference range is 550 m, the network only consists of one contention region. On the other hand, when interference range is 250 m, there are two overlapping contention regions in the network.

2.5 SUMMARY

We have shown two different approaches to describe the rate-aware resource allocation problem in multi-hop wireless ad hoc networks. As we can see, the issues of wireless nodes interference and resource allocation in wireless networks present a very different resource allocation problem as we explored in wired networks. The resource control mechanisms and policies (such as admission, scheduling, adaptation, routing) and constraints (QoS requirements and resource availability) must be considered with respect to the relations between interference and transmission ranges, rates, and capacity the wireless networks can offer. Often, the utility-based problem descriptions are considered to model resource allocations in wireless networks; however, they are complex and difficult to derive. This chapter offers also an alternative approach to consider the dual problem of resource allocation in wireless networks and look at the overall problem from the pricing point of view, which is easier to handle. In Chapter 3, we will present multiple resource allocation algorithms and QoS enforcement approaches to yield QoS guarantees such as rate, and bandwidth in wireless networks.

• • • •

CHAPTER 3
Bandwidth Management

3.1 INTRODUCTION

To achieve QoS guarantees (e.g., rate/bandwidth and jitter/end-to-end delay) in wireless 802.11-like networks is hard. Usually, these types of networks need to have in place (a) *some checking (admission) mechanisms* for constraints and computational and communication capabilities demanded by QoS requirements, (b) *resource allocation and enforcement* algorithms and protocols to deliver QoS guarantees within agreed-upon constraints, and (c) *adaptive algorithms*, mechanisms, policies, and protocols to adjust in case of unexpected bursts or changes in demands on computational and communication capabilities.

It is important to note, that even if wireless nodes perform checks on their constraints and start with resource allocations according to resource constraints discussed in Chapter 2, these 802.11-like networks do not yield strict QoS guarantees, i.e., there is no guarantee that applications get deterministically their desired QoS guarantees during their sessions. To come close to achieving at least *statistical QoS* guarantees during the runtime, one needs to have in place additional enforcement and adaptation techniques, algorithms, protocols, and services that provide (1) initial *calculation and continuous estimation* of available computational and communication resources and calculation of QoS status at times t via techniques such as *optimization solvers* for resource allocation problems (see Chapter 2.3 for problem formulations), and *heuristic and statistical estimation* approaches, and (2) *enforcement of admitted resource usage* via techniques such as *functional cross-layering between network layers* (middleware, transport, IP, and MAC layers), data *cross-layering between resources* (CPU and networks), *dynamic and integrated resource management*, and *dynamic soft-real-time resource scheduling*.

In this chapter, we explore techniques, algorithms, and protocols for *bandwidth* allocation and management in wireless 802.11-like ad hoc networks and wireless LAN networks. We will show (a) *calculations and estimation* of rate/bandwidth, (b) *cross-layer enforcement algorithms* such as rate control to achieve bandwidth guarantees, and (c) *architectures* that might be considered to achieve bandwidth resource management.

We will divide this chapter into two major sections where each section will show in a holistic manner how to calculate/estimate QoS guarantees (rate/bandwidth) as well as how to enforce them to achieve statistically strong bandwidth guarantees. In Section 3.2, we will discuss a *two-tier approximation algorithm* of the rate optimization solver. This algorithm will present the calculation of the starting rate and distribution of the rate information in wireless 802.11 multi-hop ad hoc networks. This algorithm represents actually a solution to the price-based resource allocation problem, discussed in Chapter 2.3.2. In Section 3.3, we will switch to wireless 802.11 single-hop networks and explore *dynamic bandwidth management* to provide statistical bandwidth guarantees. Each section will provide also practical comments and insights to support some of the algorithmic work. We will conclude the chapter in Section 3.4. with the summary and our lessons learned in the area of bandwidth management.

3.2 PRICE-BASED ALGORITHM FOR RATE ALLOCATION

In this section, we present a *decentralized two-tier algorithm* that represents a *solution to the price-based resource allocation problem*, formulated in Chapter 2. Recall, that we consider a wireless 802.11-like ad hoc network, where each node forwards packets to its peer nodes, and each *end-to-end flow* traverses multiple hops of wireless links from a source to a destination. The unique characteristics of multi-hop wireless networks, compared to wireline networks, is that flows compete for shared channels if they are within the interference ranges of each other. This presents the problem of designing a *topology-aware resource allocation algorithm* that is both *optimal with respect to resource utilization* and its conformation to QoS requirements and *fair* across contending multi-hop flows.

As we discussed in Chapter 2, there are several important assumptions and models under which we are considering solving the price-based resource allocation problem and designing the resource allocation algorithm to achieve optimal rate allocation:

1. we will assume the *protocol model*, especially the 802.11 MAC protocol model;
2. flows f contend for shared wireless resources in a *location-dependent manner* and create a *wireless link contention graph* G_c, depending on the wireless interference range d_{int} and transmission range d_{tx};
3. we consider complete subgraphs in G_c, called *cliques*, and especially *maximal cliques* q (it is a clique that is not contained in any other cliques) because maximal cliques represent a *basic resource unit* for defining the optimal resource allocation problem (see Figure 3.1);
4. we consider *rate* x_f of a flow and *aggregated rate* y_{ij} of all subflows along the node i and node j;
5. we describe the resource constraints on rate allocation among flows f via *clique-flow matrix* $\mathbf{R} = \{R_{qf}\}$ and it represents the resource usage pattern;
6. the overall achievable channel capacity is a vector $\mathbf{C} = (C_q, q \in Q)$ with C_q being the achievable capacity in each clique q;

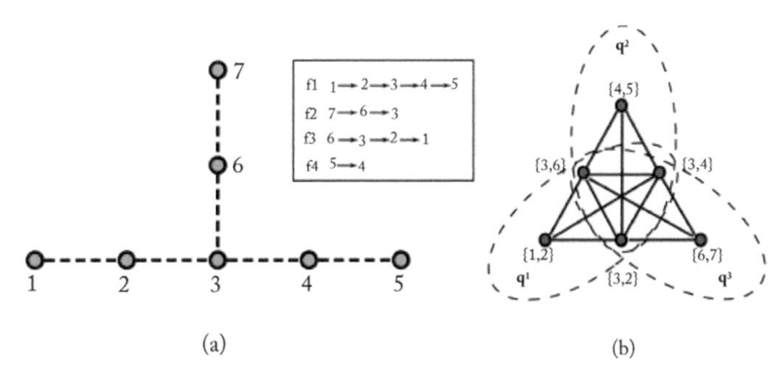

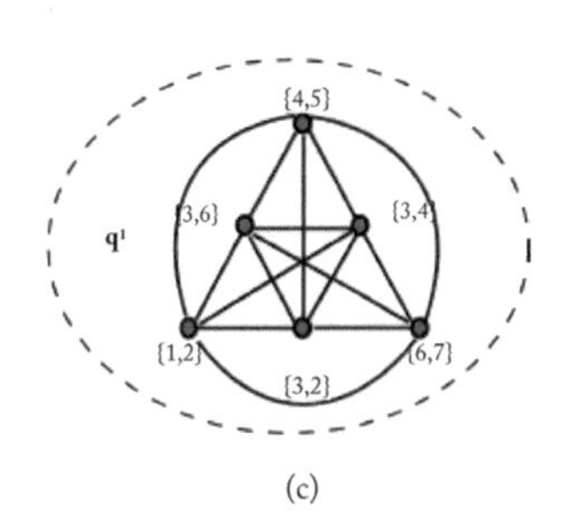

FIGURE 3.1: Wireless Link Contention Graph and Resource Constraints: Example with four flows f_1, f_2, f_3, f_4 shown in Figure 3.1(a) with network topology of wireless ad hoc nodes; and three cliques q_1, q_2, q_3 for the four flows in Figure 3.1b $d_{int} = d_{tx}$ and one clique q_1 in Figure 3.1c $d_{int} = 2^*d_{tx}$ [13].

7. we consider feasible rate allocation $\boldsymbol{x} = (x_f, f \in F)$ if, and only if, $\mathbf{Rx} \leq \mathbf{C}$;
8. we establish the *shadow price* μ_q *of a maximal clique* q, which corresponds to the Lagrange multiplier of the Lagrangian form of the utility-based resource optimization problem (see Equation 2.5 in Chapter 2);
9. the *shadow price* λ_f *of a flow* f is the sum of maximal cliques prices μ_q through which the flow f traverses, i.e., $\lambda_f = \sum \mu_q R_{qf}$ (see Equation 2.8 in Chapter 2);
10. we are considering the *price-based problem description D*, presented in Chapter 2.3.2 through equations 2.6–2.8.

Table 3.1 summarizes the symbols and model definitions from Chapter 2. Detailed summary of concepts and corresponding definitions can be found in Chapter 2, Table 2.2.

TABLE 3.1: Notations and definitions for Section 3.2.

SYMBOLS	DEFINITIONS
$G = (V, E), i \in V; e \in E;$ $E = \{e \mid e = (i,j), i \in V; j \in V, i, j$ are connected$\}$	V is set of nodes and E set of wireless links/edges , Node $i \in V$, Edge $e = (i, j)$, G is a transmission graph
d_{tx}, d_{int}	Transmission range and interference range
$f \in F; s \in S \subseteq E$	Flow f; subflow s of flow f

(*continued*)

TABLE 3.1: *(continued)*

SYMBOLS	DEFINITIONS
$G_c = (V_c, E_c)$,	Wireless link contention graph G_c with $V_c = S$
$q \in Q$	Q set of all maximal cliques in G_c
$V(q) \subseteq S$	$V(q) \subseteq S$ is the set of its vertices in contention graph G_c, where $s \in S$ is a subflow in transmission graph G
C_q	Achievable capacity at clique q
$\mathbf{R} = \{R_{qf}\}$	*clique-flow matrix* $\mathbf{R} = \{R_{qf}\}$, where $R_{qf} = \mid V(q) \cap E(f) \mid$; $\mathbf{R}$ describes the resource constraints on rate allocation among flows, and R_{qf} represents the number of subflows that a flow f has in the clique q
$\mathbf{C} = (C_q, q \in Q)$	Vector of achievable channel capacities in each of the cliques q
$\boldsymbol{x} = (x_f, f \in F)$	Feasible rate $\boldsymbol{x}$, if and only if $\mathbf{Rx} \leq \mathbf{C}$.
$U_f : \Re^+ \rightarrow \Re^+$	Utility function $U_f = \text{fun}(x_f)$ for a flow f with rate x_f contribution to utility U_f; in general, utility function represents a measure of relative satisfaction, and it is an abstract variable indicating goal-attainment or want-satisfaction
$I_f = [m_f, M_f]$	Interval I_f represents the *rate interval* from which x_f takes its values; m_f represents the minimum rate bound and M_f represents the maximum rate bound; this interval plays an important role for U_f since it assumed that over this interval I_f, the utility function U_f is an increasing, strictly concave, and twice differentiable function

TABLE 3.1: (*continued*)

SYMBOLS	DEFINITIONS
	Note: A function $f: X \rightarrow \Re$ is *(strictly) concave* if $\forall x \neq y$ and $\forall \lambda \in (0,1)$, $f(\lambda y + (1-\lambda)x) \geq (>) \lambda f(y) + (1-\lambda) f(x)$. A function f is *differentiable* at x if the derivative $f'(x)$ is defined, i.e., the graph of f has a nonvertical tangent line at a point $(x, f(x))$ and therefore cannot have a break, bend or cuts at this point. Examples of such strictly concave, increasing, and twice differentiable functions are $f(x) = \log x$ or $f(x) = \frac{x^{\alpha}}{\alpha}$ for $\alpha \neq 0,\ \alpha < 1$
$L(x;\mu) = \sum_{f \in F} (U_f(x_f) - x_f \sum_{q \in Q} \mu_q R_{qf}) + \sum_{q \in Q} \mu_q C_q$	Lagrangian form of optimization problem **P** (see Chapter 2, Equation 2.5). Presented $L(x;\mu)$ is a concrete example of a generic Lagrange function with Lagrange multipliers μ_q. Method of Lagrange multipliers provides a strategy for finding maxima and minima of a utility function U_f subject to constraints $\mathbf{Rx} \leq \mathbf{C}$
$\mu = (\mu_q, q \in Q)$	Vector of Lagrange multipliers, μ_q represents the shadow price of the clique q
$\mathbf{D}(\mu) = \max_{x_f \in I_f} L(x;\mu)$	Dual problem **D** of the primal problem **P** (see Chapter 2, Equations 2.6–2.7)
$\lambda_f = \sum_{q: E(f) \cap V(q) \neq \emptyset} \mu_q R_{qf}$	Shadow *price* of flow f
T_f	Set of time instances at which the source of flow f updates x_f
T_q	Set of time instances at which master node $v(q)$ updates μ_q

(*continued*)

TABLE 3.1: (*continued*)

SYMBOLS	DEFINITIONS
T_s^{λ}	Set of time instances at which delegation node $v(s)$ updates λ_s
T_s^{y}	Set of time instances at which delegation node $v(s)$ updates y

The design of the resource allocation solution has the objective to achieve *optimal rate assignment* with which each resource enforcement system starts to deliver desired QoS (rate) guarantee. To achieve the objective, our design will consider a two-tier resource allocation approach. In the first tier, we will design an *iterative algorithm* that determines per-clique prices μ_q and flow rate allocations x_f. In the second tier, we briefly present a *decentralized algorithm* to construct maximal cliques q. We will conclude this section with a discussion of an integration of these tiers.

3.2.1 Per-Clique Price and Flow Rate Calculation

Treating cliques as units of resource allocation, we first present an iterative algorithm that solves the problem **P** (see Chapter 2, Equations 2.2–2.4) by looking at the dual problem **D**. This iterative algorithm solves the problem **D** in two parts: (a) the *clique price update* by the clique q and (b) the *rate update* by the flow f. Clique q and f are considered as abstract entities capable of computing and communicating.

In the first part, *clique price update* process considers the following steps as time progresses, i.e., at times $t = 1,2,\ldots$

Step 1: clique q receives rates x_f from all flows f passing through clique q, where $E(f) \cap V(q) \neq \varnothing$. Note: $E(f)$ represents a set of subflows of a flow f. $V(q)$ represents a set of vertices in the contention graph G_c that belong to a maximal clique q. $V(q) \subseteq V_c$, where V_c represents all vertices in the contention graph $G_c = (V_c, E_c)$. $E(f) \cap V(q) \neq \varnothing$ means a set of subflows of a flow f that flow through a clique q, and the set is not empty.
Step 2: clique q updates clique price as follows:

$$\mu_q(t+1) = [\mu_q(t) - \gamma(C_q - \sum_{f:E(f)\cap V(q)\neq\varnothing} x_f(t)R_{qf})]^+;$$

Step 3: clique q sends clique price $\mu_q(t+1)$ to all flows f passing through clique q, where $E(f) \cap V(q) \neq \varnothing$;

In the second part, *rate update* process by a flow f considers the following steps at times $t = 1,2,\ldots.$

Step 4: flow f receives clique (channel) prices $\mu_q(t)$ from all cliques q through which it passes, where $E(f) \cap V(q) \neq \varnothing$;
Step 5: flow f calculates shadow price of a flow f: $\lambda_f(t) = \sum_{q:E(f)\cap V(q)\neq\varnothing} \mu_q(t) R_{qf}$;
Step 6: flow f adjusts rate $x_f(t+1) = x_f(\lambda_f(t))$;
Step 7: flow f sends rate updates $x_f(t+1)$ to all cliques q through which f passes, where $E(f) \cap V(q) \neq \varnothing$.

This algorithm iteratively converges to optimal (*price, rate*) allocation. The detailed proof can be found in Reference [13].

3.2.2 Decentralized Clique Construction

The first tier treats maximal cliques as entities that are able to perform communication and computation tasks. Obviously, these tasks need to be performed by the network nodes that constitute the maximal clique. As a starting point, the second tier requires a decentralized algorithm to construct maximal cliques. We will utilize the unique graphical properties of the wireless link contention graph to facilitate efficient clique construction.

We will first *decompose* the wireless ad hoc network topology into *overlapping subgraphs*, and maximal cliques are constructed based only on local topological information within each of the subgraphs. Since only wireless links that are geographically close to each other will form an edge in the wireless link contention graph, the communication and computation overhead is significantly reduced (an example of such decomposed subgraphs is in Figure 2.2, Chapter 2). It means, after we decompose wireless networks into subgraphs, maximal cliques will be constructed within the subgraphs based on local, wireless, and topological information. We will denote the *maximal clique* that contains wireless link $s \in S$ as *q(s)* and the set of all maximal cliques that contain the wireless link s as $Q(s) = \{q(s)\}$. For example, in Figure 3.1b, with $d_{\text{int}} = d_{\text{tx}}$, for a wireless link $s = \{1,2\}$, the set of maximal cliques is $Q(s) = Q(\{1,2\}) = \{q_1\}$, and for a wireless link $s = \{3,2\}$, the set of maximal cliques is $Q(\{3,2\}) = \{q_1, q_2, q_3\}$. It can be shown that the contention graph $G_c[V_c(s), E_c]$ with $V_c(s)$ being a *wireless link two-hop neighbor set* contains sufficient and necessary topological information to construct $Q(s)$ when $d_{\text{int}} = d_{\text{tx}}$; and $G_c[V_c(s), E_c]$ contains necessary topological information to construct $Q(s)$ when $d_{\text{int}} > d_{\text{tx}}$ [13].

The cliques in $Q(s)$ set are constructed by one of the vertices of s which is selected as its *delegation node*, denoted as $v(s)$. Then the delegation node $v(s)$ uses a *clique-construction algorithm*, for example, the Bierstone algorithm [16, 98] or Dharwadker's algorithm [97] to construct all maximal cliques $q \in Q(s)$ on graph $G_c[V_c(s), E_c]$. We describe the clique-construction algorithm by Dharwadker [97], using our notation in Table 3.1 and show on the example from Figure 3.1b the construction of one maximal clique q_1. One can follow the same algorithm to construct other cliques q_2 and q_3 in Figure 3.1b. Note that Dharwadker [97] also includes correctness proofs, implementation details and C++ code of the clique construction algorithm. Another software graph tool to use for implementing graph-based algorithms such as clique construction algorithms is the Graph Magics 2.1 software tool [99].

Clique Construction Algorithm [97]: Given as input the wireless link contention graph with n vertices, i.e., $|V_c| = n$, and labeled $1,2,\ldots, n$, we aim to search for a clique of size at least k. (Note that a *clique* is a complete subgraph where every two vertices in the subgraph are connected by an edge.) At each stage, if the clique obtained has size at least k, then we stop.

Part 1: For $i = 1, 2. \ldots, n$

- Initialize the clique $q_i = \{i\}$.
- Perform *procedure A* on q_i.
- For $r = 1,2,\ldots, k$ perform *procedure B* repeated r times.
- The result is a maximal clique q_i.

Part 2: For each pair of maximal cliques q_i and q_j, found in Part 1

- Initialize the clique $q_{i,j} = q_i \cap q_j$.
- Perform *procedure A* on $q_{i,j}$.
- For $r = 1,2,\ldots, k$ perform *procedure B* repeated r times.
- The result is a maximal clique $q_{i,j}$.

We will now describe Procedures A and B.

Procedure A: Given a link contention graph G_c with n vertices and a clique q of G_c, if q has no adjoinable vertices, the final output will be clique q. Else, for each adjoinable vertex v of q, find the number $\rho(q \cup \{v\})$ of adjoinable vertices of the clique $q \cup \{v\}$. Let $v_{\max}$ denote an adjoinable vertex such that $\rho(q \cup \{v_{\max}\})$ is a maximum and obtain the clique $q \cup \{v_{\max}\}$. Repeat until the clique has no adjoinable vertices.

Procedure B: Given a link contention graph G_c with n vertices and a maximal clique q of G_c, if there is no vertex v outside of q such that there is exactly one vertex w in q that is not a neighbor of v, then output q. Else, find a vertex v outside of q such that there is exactly one vertex w in q that is not a neighbor of v. Define $q^{v,w}$ by adjoining v to q and removing w from q. Perform *procedure A* on $q^{v,w}$ and output the resulting clique.

Example: Let us consider the graph in Figure 3.1b with contention graph G_c for four flows f_1, f_2, f_3, f_4, $d_{\text{int}} = d_{\text{tx}}$, and $V_c = \{v_1, v_2, \ldots, v_{|V|}\} = \{\{1,2\}, \{3,2\}, \{6,7\}, \{3,6\}, \{3,4\}, \{4,5\}\}$. We search for a clique size of $k = 4$. We will show only example for clique q_1, the other cliques' calculation is similar.

Let start with $i = 1$, which corresponds to the vertex $v_1 = \{1,2\}$ in V_c (Note that a vertex in Figure 3.1b contention graph is a link in Figure 3.1a transmission graph.), and initial value of $q_1 = \{\{1,2\}\}$. It means, the node s in V_c, $s = \{1,2\}$ is the delegation node $v(s)$ from which the clique-construction algorithm for clique q_1 starts.

Adjoinable vertex v of q_1	Adjoinable vertices of $q_1 \cup \{v\}$	$\rho\,(q_1 \cup \{v\})$
$v_1 = \{1,2\}$	$v_2 = \{3,2\}$, $v_4 = \{3,6\}$, $v_5 = \{3,4\}$	3
$v_2 = \{3,2\}$	v_1, v_4, v_5, $v_3 = \{6,7\}$	4
$v_4 = \{3,6\}$	v_1, v_2, v_3, v_5, $v_6 = \{4,5\}$	5
$v_5 = \{3,4\}$	v_1, v_2, v_3, v_4, v_6	5

Maximum $\rho(q_1 \cup \{v\}) = 5$ for v_4, so adjoin vertex v_4 to q_1. We get clique $q_1 = \{v_1, v_4\} = \{\{1,2\}, \{3,6\}\}$ with size 2.

Adjoinable vertex v of q_1	Adjoinable vertices of $q_1 \cup \{v\}$	$\rho\,(q_1 \cup \{v\})$
$v_2 = \{3,2\}$	v_3, v_5	2
$v_5 = \{3,4\}$	v_2, v_3, v_6	3

Maximum $\rho(q_1 \cup \{v\}) = 3$ for v_5, so we adjoin vertex v_5 to q_1, and we get clique $q_1 = \{v_1, v_4, v_5\} = \{\{1,2\}, \{3,6\}, \{3,4\}\}$ with size 3.

Adjoinable vertex v of q_1	Adjoinable vertices of $q_1 \cup \{v\}$	$\rho\,(q_1 \cup \{v\})$
$v_2 = \{3,2\}$	v_3	1

Maximum $\rho(q_1 \cup \{v\}) = 1$ for v_2, so we adjoin vertex v_2 to q_1, and get clique $q_1 = \{v_1, v_4, v_5, v_2\}$ at size 4, and at this point for $k = 4$, the algorithm terminates. Since there are no other vertices, we stop and obtain maximal clique q_1.

3.2.3 Two-Tier Algorithm Integration

In the first tier of the algorithm, the maximal clique q is considered as an entity that is able to perform the following tasks:

1. the clique q must *obtain the aggregated rate* $\sum_{f:E(f)\cap V(q)\neq\emptyset} x_f R_{qf}$ within it; (e.g., in Figure 3.1b, we need to obtain the aggregate rate for clique q_1 from flows f_1, f_2, f_3 that are going through the clique q_1).
2. *compute the clique-based shadow price* μ_q; i.e., we calculate the clique price according to Step 2 in Section 3.2.1.
3. *communicate the clique price* μ_q to the flows that traverse through; i.e., send the clique price to all flows that are going through the clique q (e.g., in Figure 3.1b, if we calculate the clique price for clique q_1, then the clique price must be sent to flows f_1, f_2, f_3.)

These tasks need to be distributed to the network nodes that constitute the maximal clique. If we consider the example in Figure 3.1, then network nodes from the transmission graph G that constitue the maximal clique q_1 in the link contention graph G_c will be the nodes {1, 2, 3, 4, 6}. We will present one *design choice*, where to place the tasks for clique and flow price calculation, but other implementations are possible and can be found in literature (e.g., [13]). We will allign the implementation steps with the algorithmic steps discussed in Section 3.2.1.

In our *design choice*, one delegation node in clique q serves as a *master* that performs the task of price calculation, denoted $v(q)$. For example, in Figure 3.1b, a good master delegation node for clique q_1 would be the node {3,6} in the contention graph G_c, hosted at the network node {3} in the transmission graph G. Similarly, other cliques q_2 and q_3 select their own master delegation nodes.

(Step 1) At time t, each delegation node $v(s)$ *collects the rate* of flow f which passes it (i.e., $s \in E(f)$), *computes rate* y_s at wireless link s according to $y_s = \sum_{f:s\in E(f)} x_f(t)$ and *sends it to the master nodes* $v(q)$ of all cliques q which $s \in V(q)$ belongs to. For example, in Figure 3.1b, if we consider $v(s) = \{1,2\}$ as a delegation node in V_c that collects the rates x_1 and x_3 of flows f_1 and f_3 which pass through it, then it calculates the aggregated rate $y_{12} = x_1 + x_3$, and sends it to the master node {3,6} of clique q_1 since $s = \{1,2\}$ only belongs to one clique q_1. In case of $v(s) = \{3,2\}$ in V_c, this node collects rates from flows f_1 and f_3, calculates $y_{32} = x_1 + x_3$, but then it sends it to three master nodes of cliques q_1, q_2, q_3, since this link belongs to three cliques.

(Step 2 and Step 3) The master node $v(q)$ then *computes the new price* $\mu_q(t+1)$ of clique q according to $\mu_q(t+1) = [\mu_q(t) - \gamma(C_q - \sum_{s:s\in V(q)} y_s(t))]^+$ and *distributes it to the other delegation nodes* $v(s)$ within the clique q (i.e., $s \in V(q)$). For example, in Figure 3.1b, if the master node {3,6} of clique q_1 calculates the shadow price for clique q_1, then it distributes it to all delegation nodes {1,2}, {3,2}, {3,4}, {3,6} that belong to the clique q_1. Similarly, it distributes the shadow prices for cliques q_2 and q_3 to their corresponding delegation nodes with $V(q_2) = \{\{3,6\}, \{3,2\}, \{3,4\}, \{4,5\}\}$ and $V(q_3) = \{\{3,6\}, \{3,2\}, \{3,4\}, \{6,7\}\}$, since several flows ($f_1, f_2, f_3$) pass through all three cliques.

(Step 4 and Step 5) After obtaining the updated clique price $\mu_q(t+1)$, $v(s)$ *computes a per-hop subflow price* $\lambda_s(t+1)$ according to $\lambda_s(t+1)=\sum_{q:s\in V(q)}\mu_q(t+1)R_{qs}$ for each flow f that satisfies $s\in E(f)$, then *sends* $\lambda_s(t+1)$ *to the source of f*. For example, in Figure 3.1b, the delegation node {3,2} receives the clique prices for clique q_1, q_2, and q_3 since it resides in all three maximal cliques, i.e., any subflow going through the link {3,2} will have a subflow price $\lambda_{\{3,2\}}(t+1)=\mu_{q_1}(t+1)+\mu_{q_2}(t+1)+\mu_{q_3}(t+1)$. In Figure 3.1a, flows f_1 and f_3 are going through the wireless link {3,2}, so the source node {1} for flow f_1 and the source node {6} for flow f_3 will receive the subflow price $\lambda_{\{3,2\}}$. Similarly, we calculate prices for all other subflows at each delegation node in V_c and distribute the prices for subflows to the flows' sources (nodes in V).

(Step 6 and Step 7) For the flow f, its source node performs the task of *rate update*. When the source node receives the per-hop prices $\lambda_s(t)$, it computes the *flow price* $\lambda_f(t)$ according to $\lambda_f(t)=\sum_{s:s\in E(f)}\lambda_s(t)$ and adjusts the rate x_f according to $x_f(t+1)=x_f(\lambda_f(t))$. It also notifies $v(s)$ $(s\in E(f))$ of $x_f(t+1)$. For example, in Figure 3.1a, the flow price for f_1, flowing from source node 1, through nodes 2, 3, 4, to destination node 5, will consist of per-hop subflow prices $\lambda_{\{1,2\}}$, $\lambda_{\{2,3\}}$, $\lambda_{\{3,4\}}$, $\lambda_{\{4,5\}}$. Note that the per-hop subflow prices consist of clique prices through which the subflows pass. Since subflow {1,2} passes through one clique q_1, $\lambda_{\{1,2\}}=\mu_{q_1}$. On the other hand, since the subflow {2,3} passes through three cliques, $\lambda_{\{3,2\}}(t+1)=\mu_{q_1}(t+1)+\mu_{q_2}(t+1)+\mu_{q_3}(t+1)$. Note that we are dealing with undirected graphs, hence, the price for subflow {2,3} is the same as the price for subflow {3,2}. Similar considerations are taken when determining the per-hop subflow prices $\lambda_{\{3,4\}}$, $\lambda_{\{4,5\}}$. The overall flow price for flow f_1 is then:

$$\lambda_{f_1}(t)=\lambda_{\{1,2\}}(t)+\lambda_{\{3,2\}}(t)+\lambda_{\{3,4\}}(t)+\lambda_{\{4,5\}}(t)=[\mu_{q_1}(t)]+[\mu_{q_1}(t)+\mu_{q_2}(t)+\mu_{q_3}(t)]+$$
$$[\mu_{q_1}(t)+\mu_{q_2}(t)+\mu_{q_3}(t)]+[\mu_{q_2}(t)]=3\mu_{q_1}(t)+3\mu_{q_2}(t)+2\mu_{q_3}(t).$$

Similar calculations apply to price calculations for flows f_2, f_3, and f_4 in Figure 3.1.

3.2.4 Practical Issues

There are several practical issues to be considered in realistic ad hoc network environments. First, our two-tier algorithm assumes that updates at the sources and the relaying nodes are *synchronized* to occur at times $t=1,2,\ldots$.. In realistic wireless networks, however, such *synchronization* is difficult to achieve. It means in an *asynchronous environment*, node $v(q)$ which updates the price $\mu_q(t)$ at time $t\in T_q$, (T_q is a set of time instances at which master node $v(q)$ updates μ_q.), may not have the knowledge of rate information $y_s(t)$. Instead, it only knows the sequence of recent rate updates $y_s(\tau_s^q)^1, y_s(\tau_s^q)^2,\ldots$. which satisfies $t-B\le(\tau_s^q)^1\le(\tau_s^q)^2\le\ldots\le t$, where B is the time interval when last price μ_q was calculated.

One can improve the two-tier algorithm to an *asynchronous setting*, where sending rates and clique prices are updated at different times at different nodes. For example, one approach

[13] might be for the node $v(q)$ to estimate the rate $\hat{y}_s^q(t)$ by using a *weighted average* of the received values $y_s(\tau_s^q)^1, y_s(\tau_s^q)^2, \ldots,$ and then calculate the price of the clique q according to $\mu_q(t+1) = [\mu_q(t) - \gamma(C_q - \sum_{s:s\in V(q)} \hat{y}_s^q(t))]^+$, $\forall \in T_q$. At times $t \notin T_q$, μ_q is unchanged, i.e., $\mu_q(t+1) = \mu_q(t)$. Similarly, to compute the per hop flow price $\lambda_s(t)$ at time $t \in T_s^\lambda$ (T_s^λ is the set of time instances at which delegation node $v(s)$ updates $\lambda_s(t)$.), node $v(s)$ estimates the clique price $\hat{\mu}_q^s(t)$ using a *weighted average* of the values $\mu_q(t')$ with $t - B \le t' \le t$. Then, we can calculate per hop flow price according to $\lambda_s(t+1) = \sum_{q\in V(q)} \hat{\mu}_q^s(t), \forall t \in T_s^\lambda$. At times $t \notin T_s^\lambda$, λ_s is unchanged. At time $t \in T_f$, (T_f is the set of time instances at which the source of flow f updates x_f.), the source of f estimates its flow price according to $\hat{\lambda}_f(t) = \sum_{s\in E(f)} \hat{\lambda}_s^f(t)$, where we utilize weighted average approach for estimations of $\hat{\lambda}_s^f$ and $x_f(t+1) = x_f(\hat{\lambda}_f(t))$. At time $t \in T_s^y$(T_s^y is the set of time instances at which delegation node $v(s)$ updates y_s.), node $v(s)$ calculates y_s as $y_s(t+1) = \sum_{f:s\in E(f)} \hat{x}_f^s(t)$, where $\hat{x}_f^s(t)$ is a weighted average of values $x_f(t')$ with $t' \in [t-B, t]$.

Second, realistic physical and MAC layers in wireless ad hoc networks present several challenges to deploy our price-based resource allocation algorithm:

a) The *achievable channel capacity varies* at different contention regions (cliques) depending on the MAC protocol. It is usually much smaller than the ideal channel capacity and cannot be known a priori. Dynamically estimating the achievable channel capacity (***C***) at different contention regions is a critical problem to deploy our algorithm in realistic wireless environments.
b) The two-tier decentralized clique construction and price calculation algorithm requires communication among nodes, which may introduce *additional overhead* to the network.

Third, designing an *efficient communication protocol* that still ensures appropriate algorithm convergence is also a challenging problem. We discuss briefly a possible lightweight communication architecture shown in Figure 3.2 for the two-tier resource allocation algorithm. To calculate the price of each clique, only its *gradient* (i.e., the difference between achievable capacity and traffic demand) needs to be known. Based on this observation, each wireless link calculates its local gradient by monitoring its achievable bandwidth and its traffic load (In Section 3.3., we will discuss a mechanism for achievable *bandwidth monitoring and estimation.*). Instead of communicating both load demand (flow rates x_f, aggregated rates y_s) and bandwidth information (***C***), only the gradient information is sent along with the connectivity information to construct cliques and compute their prices. To achieve low overhead communication, the information can be sent via *piggybacking*. For example, the local gradient information of each wireless link is piggybacked onto the data packets of the flows passing by to notify the downstream nodes. If we work with AODV (Ad Hoc On-

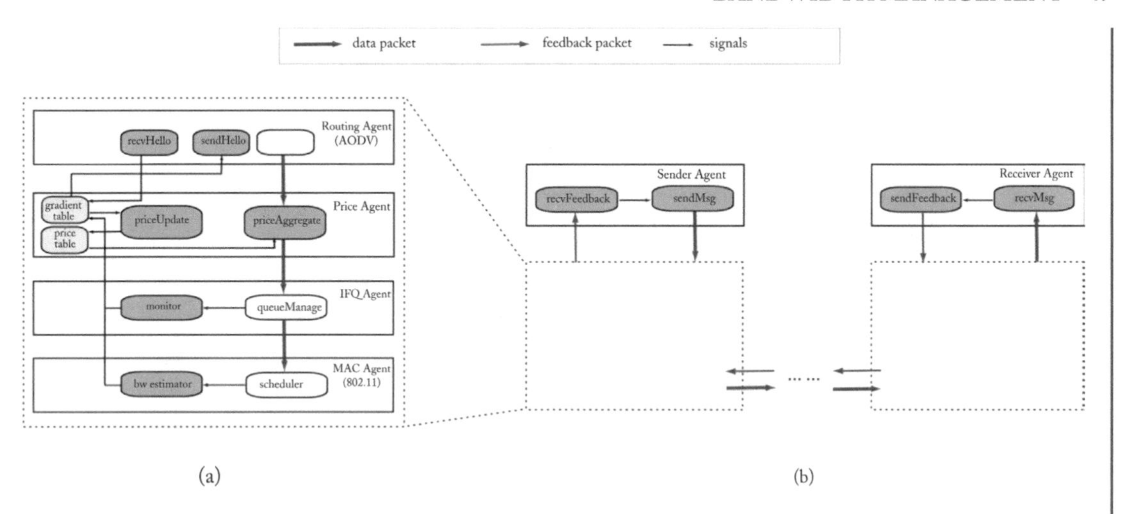

FIGURE 3.2: Implementation architecture for two-tier resource allocation algorithm.

Demand Distance Vector) routing protocol to distribute data among wireless ad hoc nodes, then we can piggyback connectivity and local gradient information onto HELLO packets and sent them at a certain time interval.

The prices can be also piggybacked onto data packets so that the destination of a flow can notify its source via FEEDBACK packets. This kind of communication architecture and protocol provide an *asynchronous information update* for price calculation and communication.

The individual components of the communication architecture in Figure 3.2a work as follows. At the MAC layer, the *bandwidth estimator* measures the local achievable bandwidth to each neighboring node. At the interface queue level, the *bandwidth monitor* observes the backlogged traffic to each neighboring node. Working with the bandwidth estimator, the monitor generates the local gradient for each wireless link to its neighbors. At the routing level, HELLO messages of the *AODV routing protocol* communicate the gradient information to its neighbors. The local gradient information, together with the gradient information received from HELLO messages, is maintained in a gradient cache table. The changes at the gradient cache table trigger the price update component, which reads the gradient information and calculates the clique prices. The price aggregation component receives data packets from the routing layer. Depending on the data packet's next hop, the price aggregation component calculates the per hop price and adds it to the aggregated price from the upstream hops. At the end nodes, as shown in Figure 3.2b, the receiving component retrieves the aggregated price information from the data packets and sends back FEEDBACK packet if it observes a price change. Upon receiving FEEDBACK packet if it observes a price change.

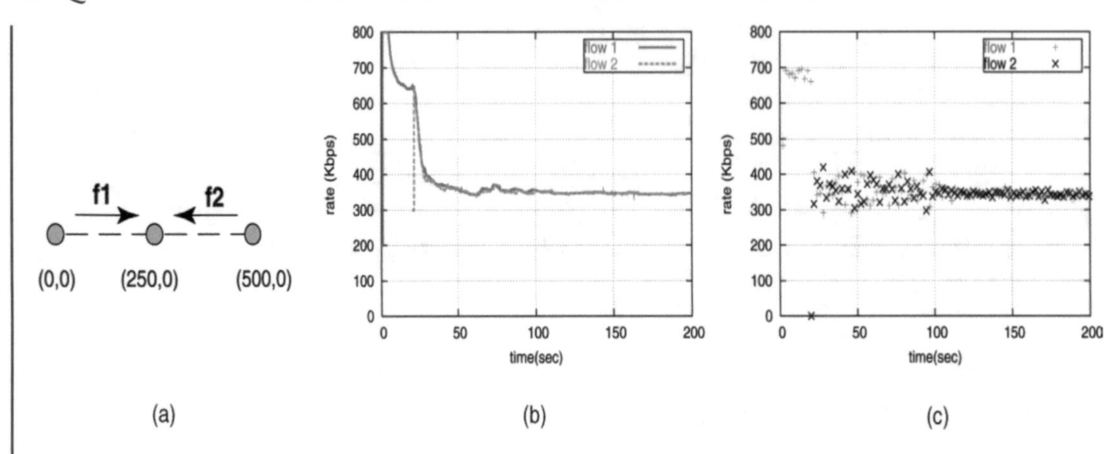

FIGURE 3.3: Convergence in hidden terminal scenario 1: (a) hidden terminal topology, (b) transmission rate, (c) throughput for two flows f_1, f_2.

Fourth, through *ns-2 simulation experiments* using realistic wireless assumptions such as the two-ray ground reflection model as the radio propagation model, 802.11 DCF as the MAC protocol, $d_{tx} = 250$ m, $d_{int} = 550$ m, data transmission rate 1 Mbps, routing protocol AODV, utility function $U_f(x_f) = \ln(x_f)$, we want to show the impact of realistic wireless interference, specifically the impact of hidden terminals in a set of special network topologies. Note that also other impacts on rate allocation need to be considered such as the impact of exposed terminals and race conditions in different network topologies [13].

Figure 3.3. shows one example of the hidden terminal scenario, as well as experimental results on the convergence of the transmission rate and the throughput of our algorithm. From the results, we observe that the algorithm performs as expected: At equilibrium, two flows share the resource fairly. The result is obvious because the sending nodes of both flows are able to obtain the information from each other, thus correctly constructing the clique and calculating its price.

In contrast, we also show the performance of the two-tier algorithm over a different hidden terminal scenario shown in Figure 3.4. In this scenario, the sending nodes of the two flows are unable to communicate, though their transmissions still interfere with each other. Thus, each wireless link treats itself as the only link within the clique, though the correct clique construction should consist of both wireless links. In this case, the price of a clique relies on the gradient of one wireless link, which in turn is calculated based on the bandwidth estimation at each node 2 or node 3. Node 2 can sense the interference from node 3, when it sends FEEDBACK packets to node 1. Similarly, node 3 can sense the interference from node 2, when it sends the data packets. But, due to their asymmetric traffic load, their bandwidth estimation results are different. As a result, the rate allocation of these two flows is not fair at equilibrium.

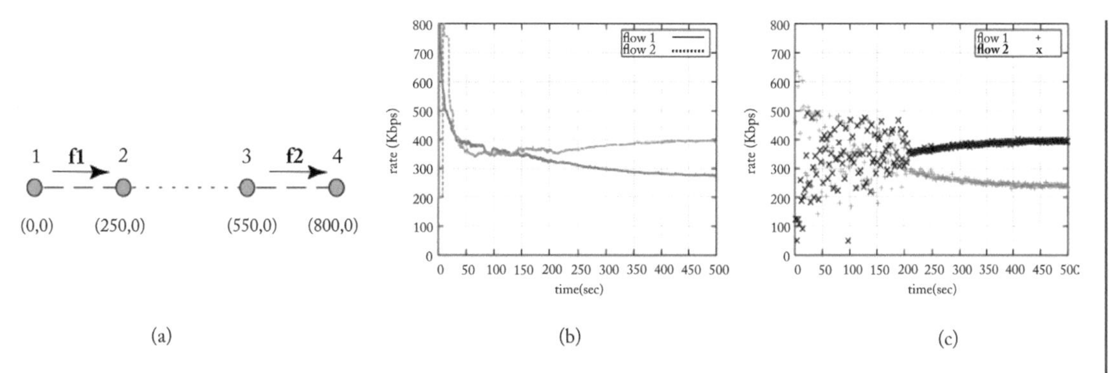

FIGURE 3.4: Convergence in hidden terminal scenario 2: (a) hidden terminal topology, (b) transmission rate, (c) throughput for two flows f_1, f_2.

3.3 DYNAMIC BANDWIDTH MANAGEMENT

As we have seen in Section 3.2, calculating and ensuring rate QoS via adaptive price/rate recalculation and reassignment can be complex in wireless multi-hop ad hoc networks. In this section, we want to show a simpler rate and bandwidth allocation approach in 802.11 wireless single hop networks. The bandwidth allocation will be simpler than what we saw in Section 3.2, but it will give us opportunity to show other resource management functions that need to accompany any resource assignment and allocation. The functions for bandwidth resource enforcement that we will discuss are *bandwidth estimation*, *bandwidth monitoring*, *bandwidth allocation*, and *adaptation*. Note that we will conduct bandwidth allocation/adaptation during the session since we are not assuming any reservation mechanisms of bandwidth. Hence, to enforce QoS such as bandwidth/rate, we need to constantly monitor, adjust, and re-allocate resources to keep QoS at statistically agreed QoS level, done by a *bandwidth management* system and via *cross-layered* approach.

We will assume that the wireless network has selected to host a *bandwidth manager (BM)* [17] which then performs the various bandwidth management functions such as the bandwidth estimation, bandwidth monitoring, and bandwidth adaptation. Figure 3.5 shows the single hop ad hoc network model as well as the bandwidth management architecture. Note that this approach works for a single hop ad hoc mode 802.11 wireless network model. The scheme can be adjusted for a single hop infrastructure mode 802.11 wireless network model with the BM residing at the access point.

Furthermore, we assume that the network has a set of flows F with $f \in F$, where each flow f is uniquely identified by its *source IP address*, *source port number*, *destination IP address*, and *destination port number*. Each new flow f registers with the BM before beginning its transmission with its bandwidth requirements $B_{\min}(f)$ and $B_{\max}(f)$. The bandwidth requirements are set in agreement

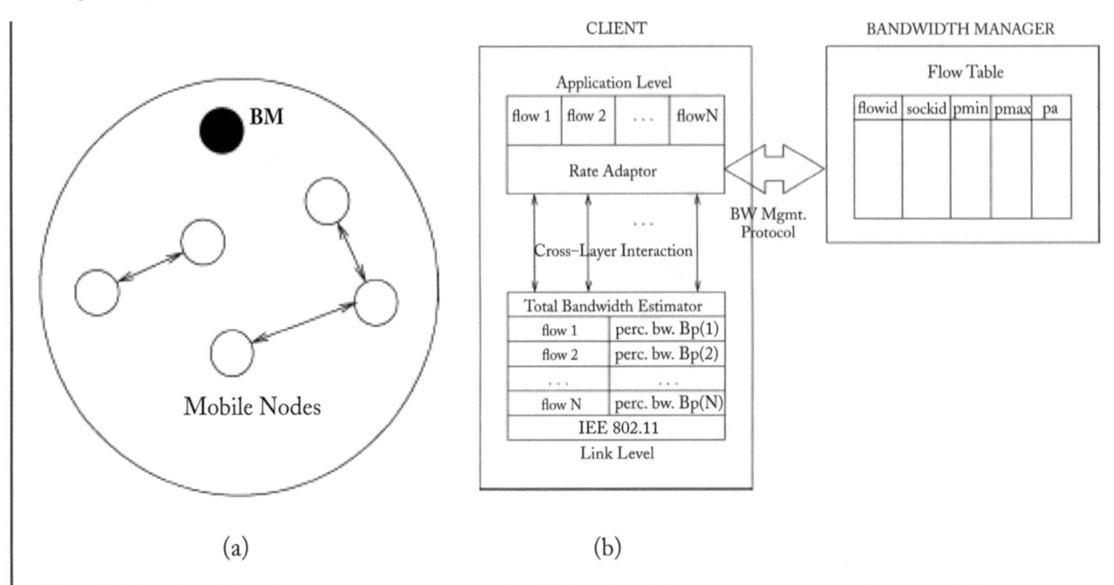

FIGURE 3.5: (a) Single hop wireless network, (b) bandwidth management architecture [9].

with the mobile application bandwidth demands. Each flow f will map its minimum and maximum bandwidth requirements to its minimum and maximum *channel time proportion* (CTP) requirements $p_{\mathrm{min}}(f)$ and $p_{\mathrm{max}}(f)$. The BM hence obtains from each flow its CTP requirements at the start of the session and gauges what proportion of unit channel time (CTP) each flow should be alloted. The CTP $p_{\mathrm{a}}(f)$ allotted by the BM (i.e., its "flow weight") lies somewhere between the flow's minimum and maximum requirements when the flow f is admitted. The term *channel time proportion* is defined as the fraction of unit time for which a flow can have the wireless channel to itself for its transmission. Since our network model allows only one node to transmit on the channel at a time, there is a direct correspondence between the channel time a flow uses and the share of the network bandwidth it receives. The BM can also refuse to admit a flow, i.e., allot 0% channel time. This can happen if the flow's minimum CTP requirement is so large that the network cannot support it, without violating some other flow's minimum CTP requirement. If the flow f is admitted, BM adds the flow f to F with its $p_{\mathrm{a}}(f)$ assignment, and the flow f uses this allotted CTP $p_{\mathrm{a}}(f)$ to calculate transmission rate. It transmits using this transmission rate until either it stops or until a new $p_{\mathrm{a}}(f)$ value is allotted to it. A new $p_{\mathrm{a}}(f)$ could be allotted to it when there is a change in the channel characteristics or in the network traffic characteristics. We assume that the flows are *well-behaved* and *cooperative*, i.e., they refrain from exceeding their alloted channel share and will release any channel share allotted to them when they stop.

In the next subsections, we discuss three parts of the bandwidth management architecture, shown in Figure 3.5 that allow us the bandwidth calculation and bandwidth enforcement: (a) the overall *BM management protocol* (in Section 3.3.1), (b) the *total bandwidth estimator*, which is needed by the BM to conduct admission control for each flow f and calculate its allotted CTP $p_a(f)$ (in Section 3.3.2), and (c) *bandwidth allocation* (in Section 3.3.3). The notation for the whole Section 3.3 is shown in Table 3.2.

TABLE 3.2: Notations and definitions for Section 3.3.

NOTATIONS	DEFINITIONS
F	Set of flows admitted by the BM
N	Number of flows in F, $N = \|F\|$
$g \in F$	All individual flows previously admitted by the BM
f	New flow requesting admission
$B_{min}(f)$	Minimum bandwidth requirement of flow f
$B_{max}(f)$	Maximum bandwidth requirement of flow f
$B_p(f)$	Total network bandwidth as perceived by flow f
$p_{min}(f)$	Minimum channel time proportion required by flow f
$p_{max}(f)$	Maximum channel time proportion required by flow f
p_{rem}	Channel time remaining after $p_{min}(g)$, $\forall g \in F$, is met
$p_{newmax}(f)$	$p_{max}(f) - p_{min}(f)$: maximum channel time proportion requirement for f that is input to max-min algorithm because $p_{min}(f)$ is already allotted
$p_{mm}(f)$	Channel time proportion allotted to flow f according to max–min fairness. This is in addition to $p_{min}(f)$, which was already allotted before max–min fairness algorithm began
$p_a(f)$	Total channel time proportion allotted to flow f, i.e., $p_{min}(f) + p_{mm}(f)$
Max–min fairness	Division of bandwidth resources is said to be *max–min fair* when: (1) the minimum data rate that a dataflow achieves is maximized; and (2) the second lowest data rate that a dataflow achieves is maximized
RTS	Request-to-Send control packet in IEEE 802.11 protocol

(*continued*)

TABLE 3.2: (*continued*)

NOTATIONS	DEFINITIONS
CTS	Clear-to-Send control packet in IEEE 802.11 protocol
ACK	Acknowledgment control packet
TP	Measured throughput
S	DATA packet size
t_s	Timestamp that the DATA packet is ready at the MAC layer
t_r	Timestamp that an ACK has been received
BW_{ch}	Channel bandwidth
T_d	Actual time for the channel to transmit DATA packet, $T_d = S/BW_{ch}$
CA	Channel allotment
R	Set of flows with $p_{min} + p_{mm}$ channel time allotment
$NEWB_p(f)$	Newly obtained bandwidth value perceived by flow f

3.3.1 Bandwidth Management Protocol

We describe the protocol used in the interaction between the various components of the bandwidth management (BM) architecture, shown in Figure 3.5b. The BM is invoked at the time of *flow establishment*, *flow teardown*, significant *change in a flow's perception* of the total bandwidth, or significant *change in a flow's traffic* pattern. The bandwidth management protocol that we discuss below is shown in Figure 3.6.

At the time of initiating a flow f (i.e., flow establishment time), the application specifies QoS requirements. It means, the application requires $B_{min}(f)$ and $B_{max}(f)$, both in bits per second to its *RA* (rate adaptor). These values have to be each divided by the flow f's perceived *total network bandwidth* $B_p(f)$ to obtain the requested CTPs $p_{min}(f) = B_{min}(f)/B_p(f)$ and $p_{max}(f) = B_{max}(f)/B_p(f)$, respectively. The $B_p(f)$ perceived by a flow f is estimated by the *TBE* (total bandwidth estimator), discussed in Section 3.3.2, at the local node. A best effort flow will have $B_{min}(f) = 0$, i.e., $p_{min}(f) = 0$. The shape of the utility curve is monotonically increasing function with increasing bandwidth availability from $B_{min}(f)$ to $B_{max}(f)$ as shown in Figure 3.7.

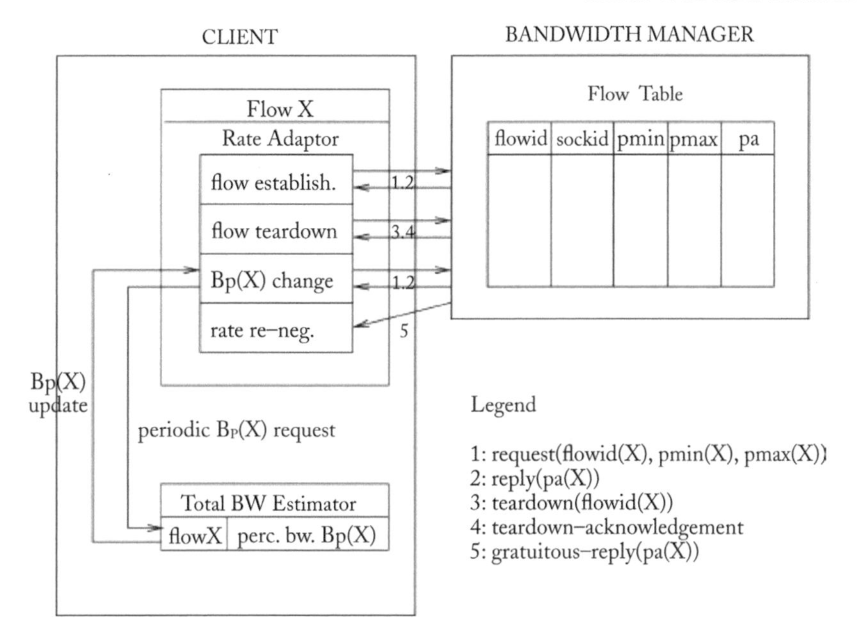

FIGURE 3.6: Bandwidth management protocol [9].

Note that both the CTP consumed by the flow *f*'s data packets in the forwarding direction as well as CTP consumed by the acknowledgements in the reverse direction, if any, must be included in *f*'s CTP requirements. Still, it is sufficient to do bandwidth estimation at only one of the end-points of the link. This is because both types of packets traverse the same wireless link and hence face the same level of contention and physical errors. The TBE (Total Bandwidth Estimation) quantifies the effect of the phenomena.

The RA of a node registers a new flow *f* with the node's TBE. Initially, the *TBE* has no estimate of $B_p(f)$ as perceived by this newly beginning flow. This is because it has to use the flow's

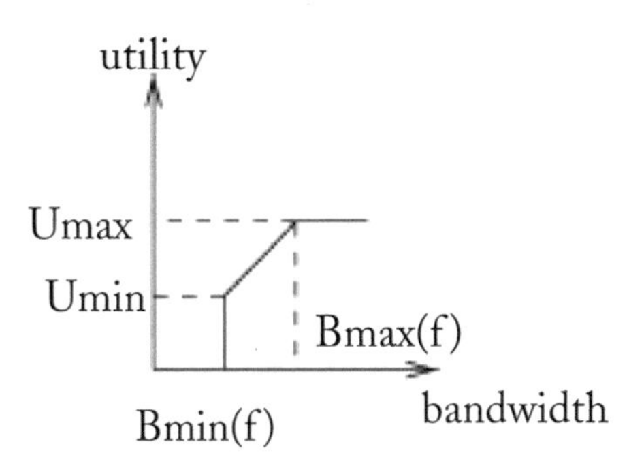

FIGURE 3.7: Utility curve of users [9].

packets themselves for obtaining an estimate of $B_p(f)$, based on the physical channel errors and contention these packets experience. But the flow has not sent out any packets yet and is still in the process of establishment. So, when initially computing the flow's requested $p_{min}(f)$ and $p_{max}(f)$, the RA has to use a hardcoded initial $B_p(f)$ estimate. Once the flow begins, a more accurate $B_p(f)$ estimation will be available from the TBE. The requested $p_{min}(f)$ and $p_{max}(f)$ CTPs can then be modified, using this new and more accurate estimate, and re-negotiation can be done with these modified values. Alternatively, in case of a connection-oriented flow, the first few flow-establishing packets can be used in the $B_p(f)$ estimation instead of the hardcoded estimate. Also, a current estimate from other flows between the same end-points can be used initially.

The RA of the new flow f sends the BM a request message containing the flow-id of f, $p_{min}(f)$, $p_{max}(f)$ and a timestamp for ordering. The BM checks whether, for all flows g in the set F of previously registered flows, it holds $1 - \sum_{g \in F} p_{min}(g) \geq p_{min}(f)$. If this is true, the new flow f is admitted ($F = F \cup \{f\}$), else it is rejected and a reply message, offering it zero CTP, is returned to its RA. Note that a best-effort flow with $p_{min}(f) = 0$ is always admitted. A rejected flow may attempt later to gain access to the channel. Flows are admitted strictly in the order they arrive to alleviate starvation of previously rejected real-time flows. The problem of starvation of a best-effort flow after admission will be discussed in Section 3.3.3.

Once the new flow f is admitted, the BM must redistribute channel time within the new set of existing flows F. Since the original admission test was passed by flow f, accommodating it will not cause the CTP allotted to any flow $g \in F$ to fall below its minimum CTP request. Hence, the BM initially sets allotted CTP $p_a(g) = p_{min}(g)$, $\forall g \in F$. The remaining channel time, $p_{rem} = 1 - \sum_{g \in F} p_{min}(g)$, is distributed among the flows $g \in F$ in max–min fair fashion. Our channel time allocation policy is therefore called *max-min fair policy with minimum guarantees* and will be discussed in Section 3.3.3.

After the new flow f is admitted, the BM registers an entry pertaining to it in its *flow table*. This entry consists of (a) the *new flow f's flow-id*, (b) the *socket description* of the socket used by the BM for communication with f's RA, (c) $p_{min}(f)$, (d) $p_{max}(f)$, and (e) $p_a(f)$. The socket descriptor is stored in the BM table so that if any re-negotiation needs to be done later with flow f's RA (e.g., when newer flows arrive in the future, or existing flows depart), this socket can be used. In addition, a *timestamp* indicating the freshness of the latest request message is also maintained for each flow. This timestamp is used for two purposes: (a) *timing out stale flow states* (reservations) and (b) *proper ordering* of multiple oustanding re-negotiation requests from the same flow. Since flow states can time-out, the entries in the flow table are *soft-state* entries. If, for some reason, a flow's reservation has timed-out, but the flow is still transmitting, this can be detected using a *policing* mechanism (see Section 3.3.3).

Finally, for every flow $g \in F$, allotted CTP $p_a(g)$ is then sent to flow g's RA using a reply message. It may be the case that not all flows $g \in F$ send a reply message. No reply message needs to be sent to a flow in F whose allotted CTP has not changed due to the arrival of the new flow f. A flow f is rejected using a unicast reply with $p_a(f) = 0$. Other existing flows' allotted CTPs are not affected.

The RA of every flow that receives a reply message from the BM sets its transmission rate to $p_a(g)*B_p(g)$ bits per second (bps), where $B_p(g)$ is the total network bandwidth perceived by flow g. The new flow f can now begin operation, whereas the older flows simply resume operation with their respective new rates.

When a flow f terminates, its RA sends a teardown message to the BM. The BM removes flow f from the set of existing flows F, i.e., $F = F - \{f\}$. It then redistributes flow f's allotted CTP $p_a(f)$ among the other flows using *max–min fair algorithm with minimum guarantees*. The RA of each flow $g \in F$ is told of its newly alloted CTP by the BM. The entry for the terminating flow f in the BM's flow table is expunged. A teardown-acknowledgment message is sent to f's RA.

Note that before we discuss in Section 3.3.3 the details of our max–min fair share algorithm with minimum guarantees, we want to present the fundamentals of the *max–min fair share algorithm* [100]. This algorithm solves the problem of dividing a scarce resource among a set of users, each of whom has an equal right to the resource, but some of whom intrinsically demand fewer resources than others. The max–min fair share algorithm intuitively allocates a fair share to a user with a "small" demand what it wants and evenly distributed unused resources to the "big" users. Formally, we define max–min share allocation to be as follows: (a) resources are allocated in order of increasing demand, (b) no source gets a resource share larger than its demand, and (c) sources with unsatisfied demands get an equal share of the resource. It means, let us consider a set of sources 1, ..., n that have resource demands $x_1, x_2, .., x_n$ (e.g., flow rate). Without loss of generality, let us order the source demands so that $x_1 \leq x_2 \leq ... \leq x_n$. Let the server be our resource and have the capacity $\boldsymbol{C}$. Then, we initially give C/n of the resource to the source with the smallest demand x_1. This may be more than what source 1 wants, perhaps, so we continue the process. The process ends when each source gets no more than what it asks for, and if source's demand was not satisfied, no less than what any other source with a higher index go.

For example, let us assume *four sources* with demands 2, 2.6, 4, and 5 Mbps, and the resource (server) capacity to be 10 Mbps. In the first round, we tentatively divide the resource into four portions of size 10/4 = 2.5 Mbps. Since this is larger than source 1's demand, we assign to source 1 2 Mbps, and this leaves 0.5 Mbps left over for the remaining three sources. We divide 0.5 Mbps evenly among the rest ($0.5/3 = 0.0\overline{6}$, giving them $2.5 + 0.0\overline{6} = 2.6\overline{6}$ Mbps each. This is larger than what source 2 wants. So we assign source 22.6 Mbps and have an access of $0.0\overline{6}$ Mbps, which we divide evenly among the remaining two sources, giving them $2.5 + 0.1\overline{6} + 0.0\overline{3} = 2.7$ Mbps each. Thus, the fair allocation is: *source 1* gets 2 Mbps, *source 2* gets 2.6 Mbps, *sources 3* and *4* get 2.7 Mbps each [101].

3.3.2 Total Bandwidth Estimation

To determine $p_{\min}(f)$ and $p_{\max}(f)$, the RA of a flow f needs to have an estimate of the total bandwidth $B_p(f)$ over the wireless link being used by the flow. We discuss a bandwidth measurement mechanism based on the IEEE 802.11, DCF (Distributed Coordinated Function) MAC (Medium Access Control) layer [9]. IEEE 802.11 relies on the DCF method to coordinate the transmission of packets based on CSMA/CA without any central control unit. We have discussed the IEEE 802.11 DCF extensively in Chapter 1, Section 1.2.1. The packet trasmission sequence and protocol are illustrated in Figure 3.8. The IEEE 802.11 protocol for single packet transmission works as follows:

Before transmitting a packet, a node senses the channel to make sure that the channel is idle; otherwise it backs off by a random interval and senses the channel again. If the channel is idle, it transmits a RTS (Request-to-Send) control packet to signal its intention to send a DATA packet. On receiving the RTS packet, the destination node replies with a CTS (Clear-to-Send) control packet to give the sender a go-ahead signal, and to silence the destination node's neighboring nodes. After receiving the CTS packet, the sender sends the DATA packet, and it is then acknowledged by an ACK control packet from the receiver.

We measure the throughput of transmitting a packet as $\mathrm{TP} = S/(t_r - t_s)$, where S is the size of the packet, t_s is the timestamp that the packet is ready at the MAC layer, and t_r is the timestamp that an ACK has been received. Note that the interval $(t_r - t_s)$ includes the channel busy and contention time. We keep separate throughput estimates to different neighboring nodes because the channel conditions may be very different. We only keep an estimate for active links, since we do not have any packets to measure $(t_r - t_s)$ over inactive ones.

This MAC layer measurement mechanism captures the effect of contention on a flow's perceived channel bandwidth. If the contention is high, $(t_r - t_s)$ will increase, and the throughput TP will decrease. This mechanism also captures the effect of physical errors because if RTS and DATA packets are affected by channel errors, they have to be retransmitted, up to retransmission limit. This increases $(t_r - t_s)$ and correspondingly decreases the flow's perceived bandwidth. It should be noted that the perceived bandwidth is measured only using successful MAC layer transmissions.

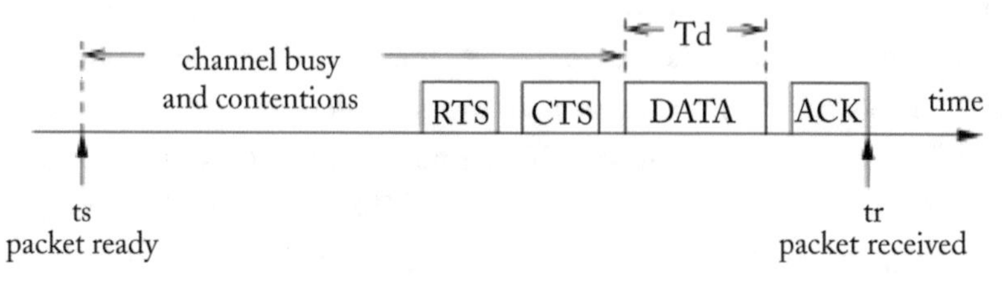

FIGURE 3.8: IEEE 802.11 protocol for single packet transmission.

It is clear that the measured throughput of a packet depends on the size of the packet. Larger packet has higher measured throughput because it sends more data once it grabs the channel. To make the throughput measurement *independent* of the packet size, we normalize the throughput of a packet to a predefined packet size. In Figure 3.8, $T_d = S/BW_{ch}$ is the actual time for the channel to transmit the DATA packet, where BW_{ch} is the channel's bit-rate. Here we assume channel's bit-rate is a predefined value. The transmission times of two packets should differ only in their times to transmit the DATA packets. Therefore, we have

$$(t_{r1} - t_{s1}) - \frac{S_1}{BW_{ch}} = (t_{r2} - t_{s2}) - \frac{S_2}{BW_{ch}} = \frac{S_2}{TP_2} - \frac{S_2}{BW_{ch}}; \tag{3.1}$$

where S_1 is the actual data packet size, and S_2 is a predefined standard packet size. By (Equation 3.1), we can calculate the normalized throughput TP_2 for the standard size packet. It should be noted that the bandwidth estimation mechanism in no way alters the IEEE 802.11 protocol. To validate the independence assumption, we show the ns-2 simulation of a group of mobile nodes within a single hop ad hoc network. We send CBR (Constant Bit Rate) traffic from one node to another and vary the packet size from small (64 bytes) to large (640 bytes) during the course of the simulation. The measured raw throughput is normalized against standard size (512 bytes). Figure 3.9 shows the result of the measured raw throughput and its corresponding normalized throughput.

As we can see, the raw throughput depends on the packet size; larger packet size leads to higher measured throughput. The normalized throughput does not depend on the data packet size. Hence, we use the normalized throughput to represent the bandwidth of a wireless link, to filter out the noise introduced by the measured raw throughput from packets of different sizes.

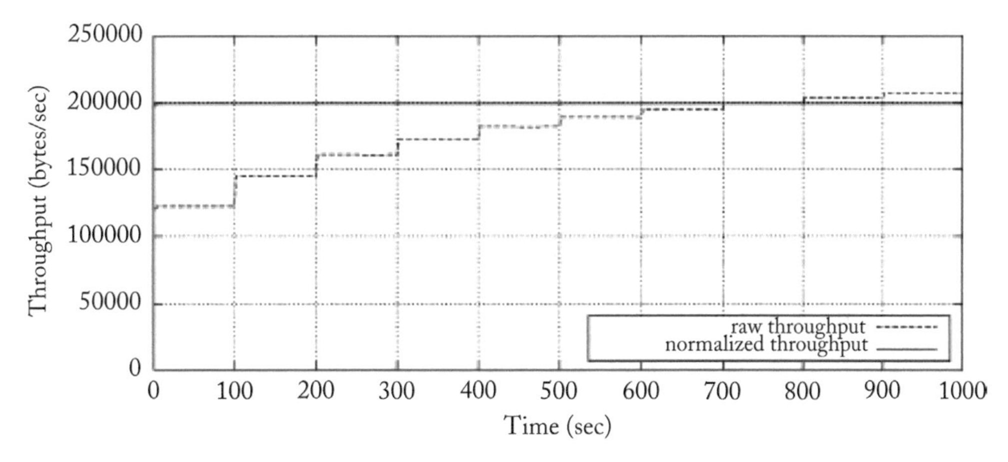

FIGURE 3.9: Raw throughput and normalized throughput at MAC layer [9].

This bandwidth estimation mechanism, with normalized extension works for many scenarios as shown in [9]. However, the theory behind the normalization may not be applicable for arbitrary large packet sizes or arbitrary high bit-error rates. In such cases, the TBE could keep an indexed table of separate estimates for different packet size ranges per active link, rather than maintaining a single normalized estimate per active link and scaling it to various packet sizes at the time of flow establishment/renegotiation. If the *indexed table estimation* method is used, the source and destination must both perform total bandwidth estimation for data and acknowledgments, respectively. The destination must periodically communicate its bandwidth estimate for acknowledgment packets with the source using an *in-band signaling* mechanism. In the single normalized estimate method, the source alone does the estimation and appropriately scales the normalized estimate for both data and acknowledgement packet sizes. Thus, although the indexed table estimation method improves accuracy of the estimate in certain special cases, it also incurs a small storage space and in-band signaling overhead.

3.3.3 Bandwidth Allocation and Adaptation

Fairness is an important issue in designing the bandwidth allocation within the BM. We describe the *max–min fairness algorithm with minimum guarantees* in allotting channel time to the flows.

Recall that in *max–min fairness* [e.g., 17, 100], flows with small channel time requests are granted their requests first; the remaining channel capacity is then evenly divided among the more demanding flows. In our bandwidth allocation case, $p_a(f)$ is first set to $p_{min}(f)$ for all the flows. The channel time that remains, p_{rem}, after satisfying the flows' minimum requirements, is allotted to the flows in *max–min fashion*. The new maximum requirement for each flow in the max–min algorithm is

$$p_{newmax}(f) = p_{max}(f) - p_{min}(f),$$

because $p_{min}(f)$ has already been allotted to it and must be subtracted from the original maximum requirement. We denote the channel time, allotted to flow f by the max–min algorithm, as $p_{mm}(f)$. This is in addition to $p_{min}(f)$ allotted before the max–min algorithm is even invoked. The computation of the max–min allocation is as follows:

- Initially, the set of flows f, whose new maximum channel time requirement $p_{newmax}(f)$ has been satisfied, is empty: $\boldsymbol{R} = \varnothing$.
- Next, we compute the first-level allottment as $CA_0 = p_{rem}/N$, where N is the total number of flows.
- Next, we include all flows f with $p_{newmax}(f) < CA_0$ in set $\boldsymbol{R}$, and allot each of them $p_{mm}(f) = p_{newmax}(f)$.

- Next, we compute $CA_1 = (p_{\text{rem}} - \sum_{f \in R} p_{\text{newmax}}(f))/(N - |R|)$.
- If for all flows g $\notin R$, $p_{\text{newmax}}(g) \geq CA_1$, then we allot each of them $p_{\text{mm}}(g) = CA_1$ and stop.
- Otherwise, we include those flows g with $p_{\text{newmax}}(g) < CA_1$ in set $\boldsymbol{R}$, allot each of them $p_{\text{mm}}(g) = p_{\text{newmax}}(g)$, and recompute the next level CA_2.

When the algorithm terminates, the allocation $p_{\text{mm}}(f)$ for all flows is *max–min fair*. The computational complexity of this algorithm is $O(N^2)$. Also, after every flow f's, $p_{\text{mm}}(f)$ has been determined using max–min algorithm, the BM sets $p_{\text{a}}(f) = p_{\text{min}}(f) + p_{\text{mm}}(f)$ and returns this value to flow f's RA. Figure 3.10 shows the max–min fair resource allocation algorithm.

As we mentioned in Section 3.3.1, *best effort flows* are only given access to the channel after all real-time flows' minimum guarantees are satisfied. This could lead to *starvation of best-effort flows* in case $\sum_{g \in F} p_{\text{min}}(g) \rightarrow 100\%$. One way to eliminate this problem would be to partition the channel

```
Input. Channel time: p_rem; set of requests: p_newmax[f]
Output. Set of allocations: p_mm[f]
proc Max-min(p_rem, p_newmax[f]) ≡
  R := {}; //set of satisfied flows
  N := size_of(p_newmax[f]);
  p_mm[f] := 0;
  while (true) do
    total_satisfied = 0;
    foreach f ∈ R do
      total_satisfied+ = p_mm[f];
    od
    CA := (p_rem − total_satisfied)/(N − size_of(R));
    stop := true;
    foreach f ∉ R do
      if (p_newmax[f] < CA) then
      R := R + {f};
      p_mm[f] := p_newmax[f];
      stop := false;
      fi
    od
    if (stop) then
    foreach f ∉ R do
    p_mm[f] := CA;
    od
    break;
    fi
  od
```

FIGURE 3.10: Max–min fair share resource allocation algorithm [9].

time into a *large minimum guarantee portion* and a *small max–min fair portion* (similar to bandwidth partitioning in Reference [18]). The minimum requirements of real-time flows will be allotted only from the minimum-guarantee portion. The max–min portion, along with any left over minimum-guarantee portion, is used to allot the flows' extra CTP $p_{\mathrm{mm}}(g)$, using just a max–min scheme. Both real-time and best effort flows with $p_{\mathrm{newmax}} > 0$ can compete for this portion, and best effort flows will not starve. The disadvantage of having separate max–min fair portion is that the channel time available to satisfy minimum guarantees of real-time flows is reduced, and so we could see more real-time flows dropped.

Another alternate scheme involves *pricing of channel time* and enforcing priorities based on *flow budgets*. The max–min fair policy with minimum guarantee lends itself to an elegant *two-level pricing scheme*. The first level represents the guaranteed minimum CTP $p_{\mathrm{min}}(g)$ which is valued at a *substantial price*, whereas at the second level, a channel time $p_{\mathrm{mm}}(g)$ is *relatively very cheap*. Under this two-level pricing scheme, users would be inclined to request as little minimum guaranteed bandwidth as possible, in order to save cost. It means minimum requirements are thus "punished," while high maximum requirements carry no penalty. The BM adjusts the price so as to trade-off *blocking probability* of the flows with its revenue. If the price is too high, too few flows can afford it, and hence, blocking probability is high. If the price is low, blocking probability is low, but revenue may suffer. Pricing for wireless networks has been studied previously [13, 19, 20, 21], but the above two-level approach is especially suitable for the bandwidth allocation policy, discussed in this section. Note that in Section 3.2, we have discussed flow pricing mechanims as well, but with *one price for the calculated and assigned flow rate* (no varying prices for rates assigned according to max–min fair share).

In Section 3.3.1, we have discussed that the bandwidth management scheme includes *policing mechanism*. Policing refers to the task of monitoring users to make sure that they conform to their allocated banwidth. The BM operates in two modes: *normal* and *policing*. When operating in *policing mode*, the BM listens promiscuously to the network traffic and checks whether a flow, identified by the source and destination addresses and port numbers in its packet headers, is sending out packets faster than its allotted rate. Additionally, it can catch those flows who have not registered with the BM. This can be some type of "denial-of-service attack" by a malicious user or it can be caused by some unmanaged applications.

Operating in policing mode is expensive. Therefore, the BM should operate in this mode only when necessary. To this end, the BM relies on the sudden, sharp decrease of channel bandwidth as an indication, in the renegotiation process. If there is a sudden flock of renegotiation requests due to reduction of $B_{\mathrm{p}}(g)$, it is likely that abnormally high channel contention has occurred. Subsequently, the BM switches into policing mode to monitor the activity of the network. It may be that the channel contention is due to a sudden increase in physical errors, or it may be that it is due to a malicious or unmanaged flow. The policing scheme can identify which of the above is the cause. It could also happen that the unreliability subnet broadcast reply message did not reach a particular RA, so a flow is continuing to transmit packets faster than its re-allotted rate.

As new flows arrive and existing flows leave, *change in a flow's perception of the total network bandwidth* happens. The RA of every flow periodically obtains from the TBE the flow's current perceived total bandwidth. The TBE updates the RA with the mean of the perceived total network bandwidth measured for each packet successfully transmitted by the flow in recent history. The inter-update period could be in terms of number of packets transmitted or in terms of time. We recommend using a *hybrid scheme* for determining update period: it should be based on time when the transmission rate of the flow is low and based on number of packets transmitted when it is high.

In case a newly obtained perceived bandwidth value $\mathrm{NEWB_p}(f)$ differs significantly from $B_\mathrm{p}(f)$, the RA must renegotiate its flow's CTP with the BM. It must also set the value of perceived bandwidth $B_\mathrm{p}(f)$ to the newly obtained value $\mathrm{NEWB_p}(f)$. Note that the RA only sets $B_\mathrm{p}(f)$ to $\mathrm{NEWB_p}(f)$ and renegotiates with the BM using this new value when there is a significant change, not with every update. A new rate using the previous allotted CTP is, however, calculated with every update. If renegotiation has to be done, the RA of the flow f sends a request message to the BM with *flow-id*, $p_\mathrm{min}(f)$ and $p_\mathrm{max}(f)$. The values of $p_\mathrm{min}(f)$ and $p_\mathrm{max}(f)$ sent in the request message are recalculated using the new value of $B_\mathrm{p}(f)$. The rest of the renegotiation procedure is almost identical to the one used for flow establishment in Section 3.3.1, both at the BM and at the RA. The only difference is that the BM does not have to add new entry into its flow table for f. Note that a flow f's renegotation request can be rejected by the BM in response to the requested CTP. This means that the flow has been cut-off in mid-operation. Unfortunately, the nature of the wireless network is inherently unreliable, and as network resources decrease, some flows will necessarily have to be cut-off in mid-operation so that others can be supported, but those that are supported get them $p_\mathrm{min}(f)$ CTP.

The BM must also deal with *change in a flow's traffic characteristics*. For example, when a VBR (variable bit rate) UDP (user datagram protocol) flow f (e.g., a flow carrying MPEG video compressed traffic) needs to send a burst of traffic at a rate different from its normal rate, it must inform its RA. The RA will renegotiate for a larger CTP for flow f depending on the bandwidth of the burst. The renegotiation procedure is the same as in the case of change in perceived bandwidth. At the end of the burst duration, the RA will again renegotiate to release the excess CTP. This solution is equivalent to splitting up a VBR stream in the time domain into multiple CBR streams [22]. Frequent bursts could result in an explosion in renegotiation overheads. One can deal with the problem of frequent bursts as follows: (a) setting $B_\mathrm{min}(f)$, at the time of burst-induced renegotiation, large enough to engulf multiple bursts and (b) having large buffering at the receiver to deal with the burst.

3.3.4 Practical Issues

When deploying the admission control and the dynamic bandwidth management system in 802.11, several practical issues need to be considered.

First, one needs to evaluate the 802.11 wireless system with BM in terms of (a) *number of flows* we can actually support and (b) *minimal and fair bandwidth* we can achieve for all admitted flows. We will show selective results using the ns-2 simulation evaluations [9] with IEEE 802.11 DCF protocol, 20-node network, 170 m × 170 m network area, 10 flows, maximum bandwidth requirement of 200 kbps (typical of an audio streaming application), minimum bandwidth requirement of 100 kbps, 512 bytes per packet, RA's inter-update interval 100 packets, and its perceived bandwidth variation-tolerance threshold $\delta = 15\%$, by default. In Figure 3.11, we compare throughput (bandwidth) performance of two IEEE 802.11 systems, the *base IEEE 802.11 system* without BM and admission (see Figure 3.11a), and the *enhanced 802.11 system* with BM and admission control (see Figure 3.11b). We want to stress two important issues, the *throughput fairness* when supporting 10 flows and the provisioning of minimal bandwidth requirement for flows in the system. Figure 3.11a shows clearly chaotic plot with many flows not being treated fairly and flows often falling far below their minimum bandwidth requirement (24 packets per second, shown with the horizontal line). In Figure 3.11b, the results show the gradual adjustment of the throughput as new flows join the system. When a new flow is admitted, contention increases for all the existing flows, and the allotted bandwidth gets readjusted. However, as we can see, most of the flows stay above their minimum bandwidth requirement. The BM scheme hence ensures that the bandwidth is allocated according to max–min fairness, and the *minimum bandwidth requirements* of the flows are met with a far higher probability than the base IEEE 802.11 scheme. Furthermore, the enhanced scheme reduces *jitter in throughput* compared to base IEEE 802.11, where throughput jitter is the difference in throughput observed over two consecutive same-sized time intervals.

Second, under dynamic bandwidth management, if new flows are being admitted, contention increases for all existing flows. One needs to have *flow dropping strategies* available to respond to contention situation. In general, the flow that notices an "unacceptably" poor channel quality and "complains" first is dropped. Another strategy can be to drop the flow last admitted. Pricing could also play a role here: the flow paying the least can be dropped.

Third, flows might have *different minimum bandwidth requirements* and *varying arrival times* which is challenging to satisfy. Figure 3.12 shows the comparison between based and enhanced IEEE 802.11 networks with six-node wireless network, 100 packets default inter-update interval, three UDP flows with minimum requirements of 100 , 200, and 400 kbps, respectively, and with maximum requirement of 600 kbps for all three flows. As we can see in Figure 3.12a, the base 802.11 case *cannot differentiate* among the throughput demands, i.e., if there is bandwidth in the network, all three flows will go with maximum requirement. On the other hand, Figure 3.12b shows

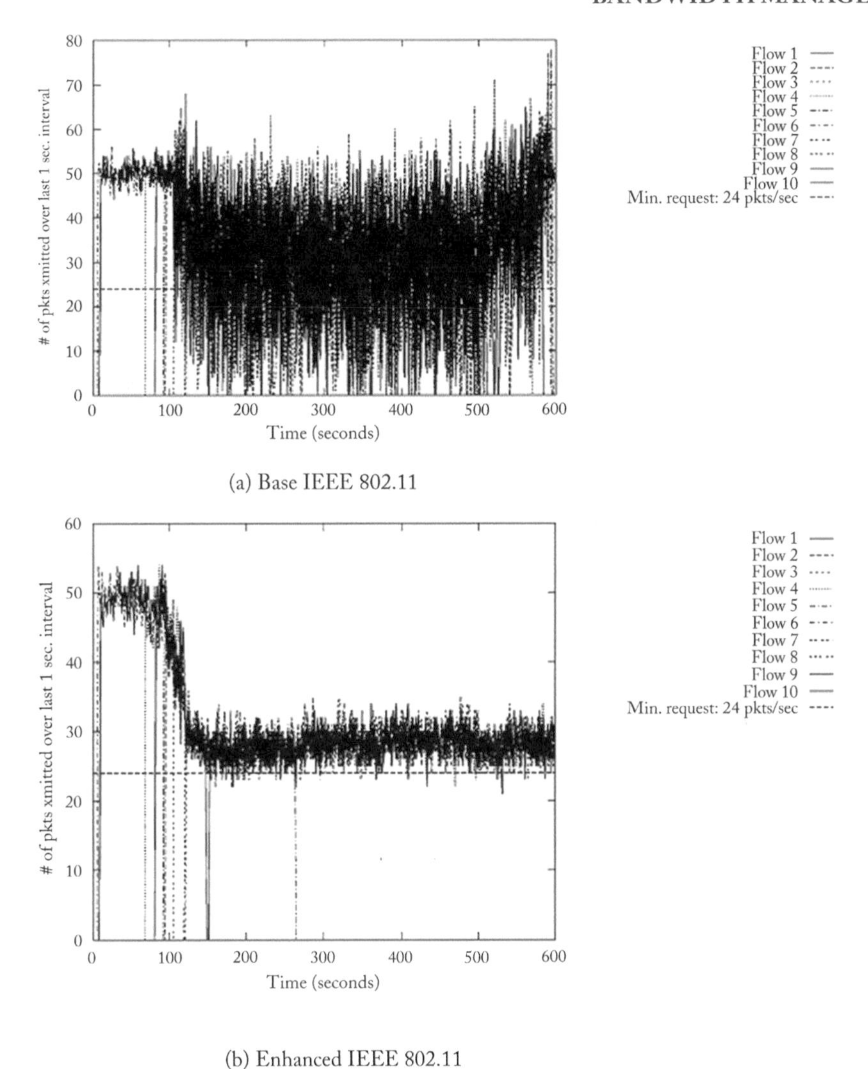

FIGURE 3.11: Comparative throughput performance of base and enhanced IEEE 802.11 network for 10-flow scenario [9].

enhanced 802.11 case which *differentiates among the three flows* without violating the minimum requirements for each flow. In Figure 3.12c, we show a three-flow scenario with each flow having a minimum requirement of 200 kbps and maximum requirement of 600 kbps. The enhanced 802.11 case *adjusts the throughput of flows* as new flows' start time is staggered and the flows are arriving at different start times without violating minimum bandwidth requirement.

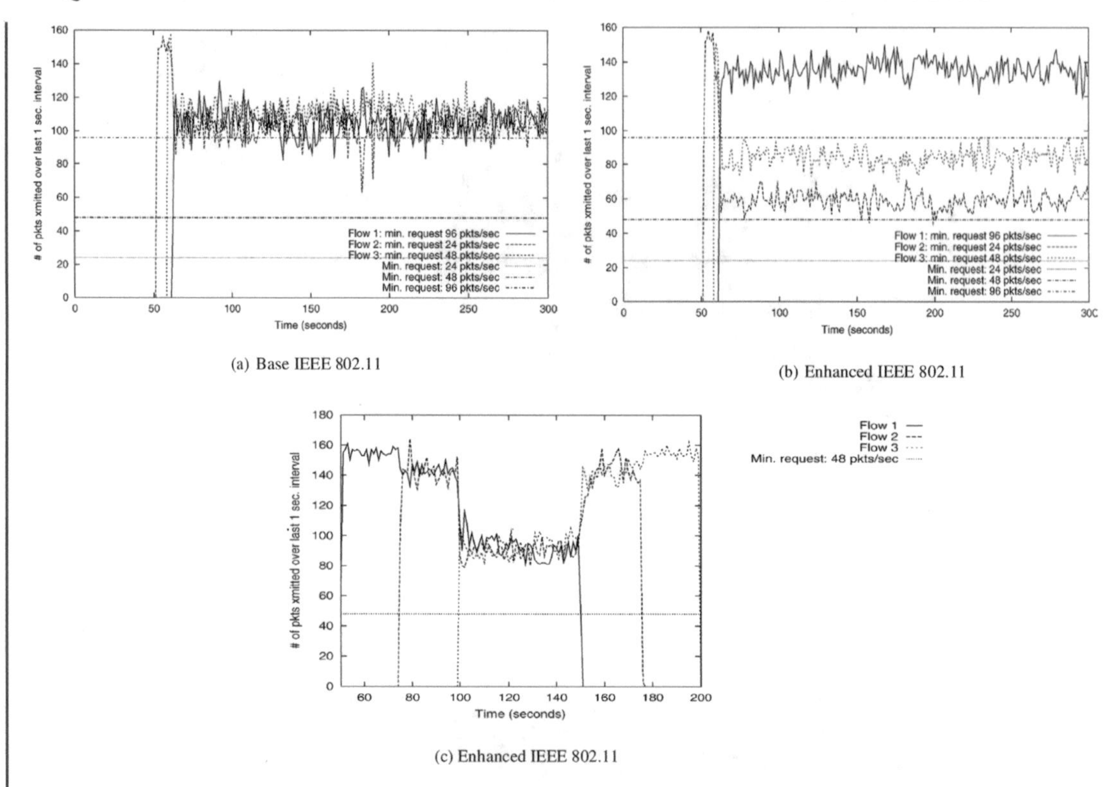

FIGURE 3.12: (a) and (b) Comparative throughput performance of base and enhanced IEEE 802.11 for three-flow scenario with different minimum bandwidth requirements; (c) Enhanced IEEE 802.11 throughput performance for three-flow scenario with staggered start times [9].

Fourth, if we consider TCP traffic over IEEE 802.11, the question is what is the impact of BM on TCP performance in terms of BM placement and sizes of queues in the protocol stack. Note that TPC traffic is a best effort and elastic traffic with $p_{\min}(f) = 0, p_{\max}(f) = 100\%$. While in the UDP experiments, usually the BM rate control using the RA is done *in the UPD application*; in the TCP experiments, queue-based BM rate control is done per node at the *network interface queue* (below TCP transport layer) The network interface queue only releases packets at the rate allotted by the BM. Under this BM rate control placement, there are then two cases to consider: (a) *network interface queue size is smaller* than the maximum congestion window size for a TCP flow, (b) *network interface queue size is larger* than the maximum congestion window size for a TCP flow.

In the (a) case, if we have BM in place (enhanced 802.11 scheme), a TCP flow will keep increasing its congestion window size up to the queuing limit as shown in Figure 3.13 (b). Without BM in place (base 802.11 scheme), a TCP flow has to fill the router queue before it cuts back its

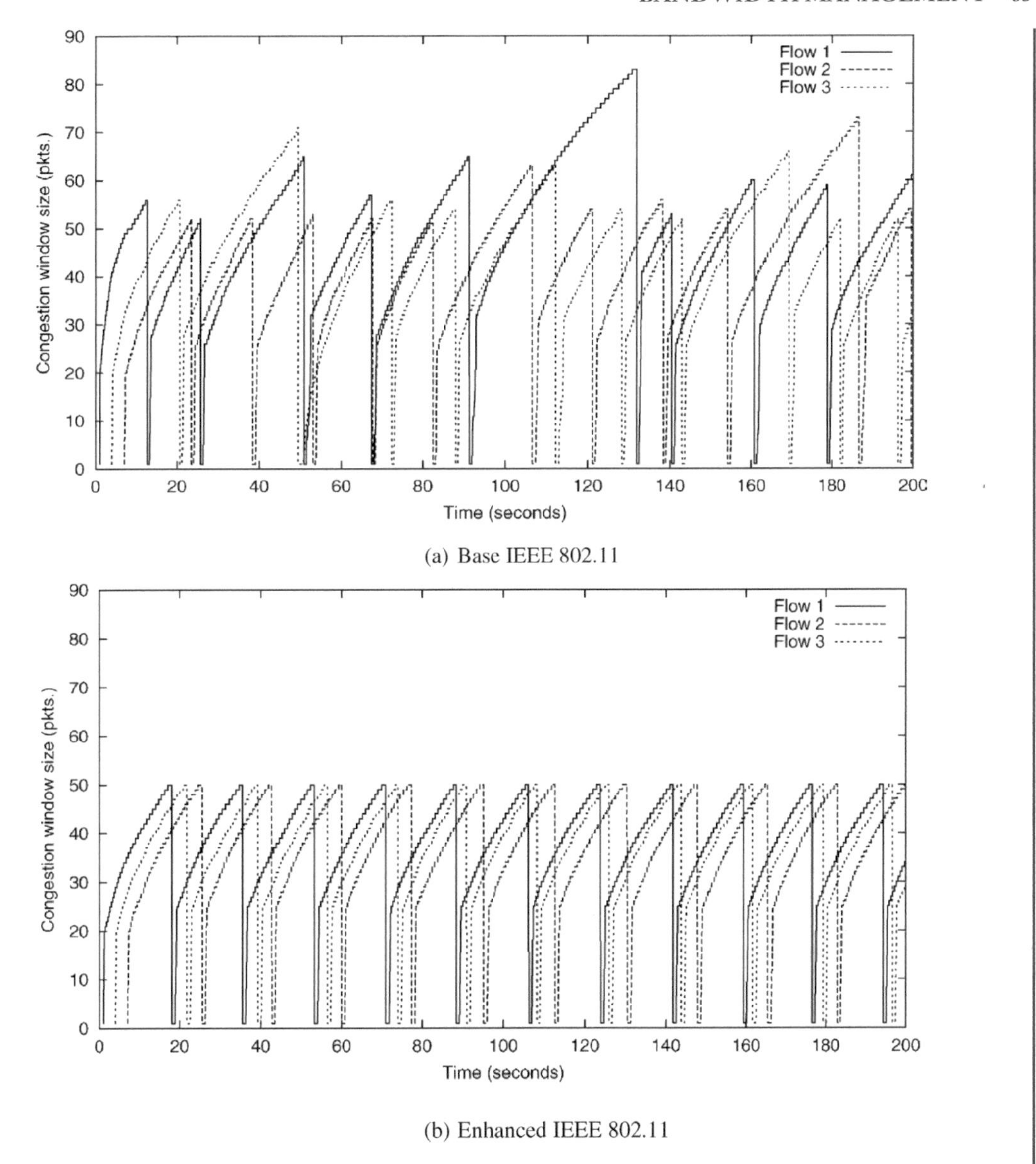

(a) Base IEEE 802.11

(b) Enhanced IEEE 802.11

FIGURE 3.13: Comparison of TCP congestion window behavior when interface queue size is smaller than congestion window (network interface queue is 50 packets, maximum TCP congestion window size is 128 packets) [9].

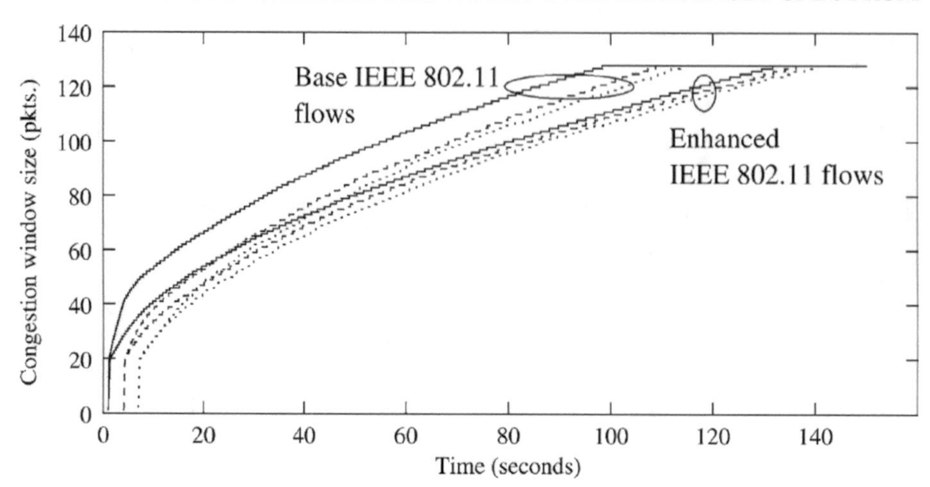

FIGURE 3.14: Comparison of TCP congestion window behavior when interface queue size is larger than congestion window limit [9].

congestion window size which incurs unnecessary long queuing delay for the packets. The unmanaged release of packets from the queue (base case) results in unequal congestion window growth and causes unfairness as shown in Figure 3.13 (a). As a result, the fairness and jitter of the flows deteriorate, and the number of dropped packets is significantly larger than the enhanced 802.11 scheme. The reason for larger number of dropped packets in the base case is because an entire window of packets may be dropped at a time before TCP resets its congestion window size, whereas in the enhanced scheme, a single packet loss results in window reset. However, this also results in smaller throughput for the enhanced scheme than for the base scheme.

In the (b) case with BM in place, the TCP's congestion window size can never reach the maximum network interface queue size, and hence, there is no packet loss as a result of queue overflow as shown in Figure 3.14. In this case, we can expect TCP's congestion window size to stay at its maximum limit without fluctuation because there is no packet loss at the MAC layer either. Note that the slow convergence speed of TCP's congestion window size does not impact its throughput efficiency, as the interface queue is kept non-empty all the times. However, in order to minimize queueing delay, it is advisable to set TCP's congestion window limit to a small value when running over a bandwidth managed network.

3.4 SUMMARY

In this chapter, we have shown different types of resource allocation algorithms for multi-hop ad hoc wireless networks as well as for single hop ad hoc wireless networks. The bandwidth/rate allocation for multi-hop ad hoc wireless networks needs to take into account the interference clique-flow information to come up with appropriate rate allocations for each flow. We have shown in this space a *price-based distributed rate allocation algorithm* that converges and is feasible, as well as one possible two-tier design of this algorithm. As the Section 3.2 showed the multi-hop resource allocation is complex, and if end-to-end QoS guarantees want to be achieved, careful consideration of nodes neighborhood must be considered.

In Section 3.3, we have shown a simpler resource allocation technique for statistical bandwidth guarantees over single hop wireless networks. In addition to the resource allocation, we have also discussed the corresponding architecture, protocol, and services that need to accompany bandwidth resource allocation. As we can see, the bandwidth resource allocation approaches are accompanied with flow admission function, enforcement algorithms, and strong monitoring capabilities to estimate and predict the rate, channel time, and bandwidth, in order to account for the wireless contention and physical errors on the shared wireless channel. In Chapter 4, we will concentrate on the delay QoS and resource approaches to deliver delay-sensitive bounds in wireless networks.

• • • •

CHAPTER 4

Delay Management

4.1 INTRODUCTION

Owing to the proliferation of mobile devices such as smart phones and of static smart sensors such as power grid's PMU (phasor measurement units), which can run in IEEE 802.11 wireless infrastructure and ad hoc modes, it is becoming increasingly important to consider QoS delay provision for multimedia applications such as real-time audio/video applications on the WiFi-enabled phones, as well as sensory applications such as SCADA (Supervisory Control and Data Acquisition) applications on WiFi-enabled PMUs and their collecting gateways. The problem is that even if we consider QoS-aware mechanisms and algorithms at lower MAC/network layers as discussed in Chapter 3, it might not be sufficient to provide *end-to-end delay guarantees* for these applications. Therefore, we need to adopt two important mechanisms for end-to-end delay provisioning: (a) *upper layer adaptation mechanisms* utilizing information from delay-sensitive applications to achieve more flexibility and get better application feedback about data differentiation and (b) *cross-layer integration* between upper layer (application and middleware) and lower layer (MAC, network, and OS) delay-aware algorithms.

A delay-sensitive application usually has QoS requirements on two quantifiable metrics, *end-to-end latency* and its *variance (jitter)*. Depending on the network conditions, applications may have *multilevel QoS requirements*. For example, when we make a phone call using VoIP (Voice Over IP), we prefer to have *acceptable voice quality* (quality level 1) rather than drop the call if the network condition is bad. On the other hand, if the network condition is good, we prefer *high voice quality* (quality level 2). In applications such as VoIP to achieve appropriate quality, we present *application layer adaptation* and corresponding *cross-layering* architecture (see Section 4.2) to deliver end-to-end delay in a range of one of the desired delay quality levels, depending on network conditions.

On the other hand, if we consider applications such as SCADA with PMU sensors, measuring voltage of transmission power lines, and sending measurements over wireless networks to collection gateways, we have tighter deadlines and stricter delays than we have in the VoIP example (see Section 4.3).

Before we continue with the discussion on individual delay management techniques, we want to stress couple of points that will not be presented in depth in the following subsections:

1. *Upper layer adaptation as well as system adaptation* are very important concepts in mobile and wireless computing and have a broad applicability. Especially, if we deal with mobility of users and their phones, mobile elements must be adaptive [103, 106]. As shown in Reference [103], the range of adaptive strategies is deliminatted by two extremes. At one extreme, adaptation is embedded only in the application, called *application-only adaptation*, and the application takes full responsibility for adaptation. While this approach avoids the need for any system support, it makes the programming of an application very complex and expensive. The other extreme includes *system-only adaptation*, also called *application-transparent adaptation*, where the responsibility for adaptation is carried by the underlying system. While this approach provides backward compatibility with existing applications, it also may provide adaptation inadequate or even counterproductive for some applications. Hence, we will explore *adaptive QoS solutions* for IEEE 802.11 applications/systems between these two extremes, collectively called *application-aware adaptations.* In case of application-aware adaptations, applications and underlying system/network layers collaborate and adapt together. It means, applications are able to determine how best to adapt, and the system is able to monitor resources and enforce allocation decisions [103]. Note that one of the first systems using the application-aware adaptations was the Odyssey platform for mobile computing [104]. An example of a system with application-transparent adaptation is the system Puppeteer [105], which supports adaptation of component-based applications in mobile environments. It takes advantage of the structured nature of documents the applications manipulate to perform adaptation without modifying the applications.
2. The *adaptation flexibility* very much depends on the application demands and end-to-end delay requirements. For example, in case of a VoIP application, we have much more flexibility to adapt and adjust to different delay quality levels than in the SCADA application. Hence, the tighter the delay QoS requirements are, the less mobility and adaptation are allowed for QoS provisioning. For example, in case of SCADA, the PMU sensors are static and are tightly scheduled as we show in Section 4.3. On the other hand, in case of VoIP, if users with phones move, adaptive levels of delay QoS in IEEE 802.11 wireless networks must be present.
3. In this lecture, we focus on bandwidth and delay QoS provisioning over IEEE 802.11 wireless networks. However, the reader should be aware that also other resources such as *energy* can be treated along these lines, even though this lecture focuses only on the two QoS metrics [107]. Furthermore, cross-layer approaches with application-aware adaptation can optimize joint bandwidth, delay, and energy QoS metrics in mobile devices as shown, for example, in the GRACE-1, GRACE-2, or DYNAMO systems [108, 109, 110] to support multimedia applications.

4. In this chapter, we address the *delay control problem* and not the bandwidth utilization problem as discussed in previous Chapters 2 and 3. It means we do not show the increase in the total bandwidth utilization, but rather, we show under the same utilization level that a delay-sensitive traffic will be able to gain more bandwidth, therefore smaller delays.

The outline of this chapter is as follows. In Section 4.2, we discuss QoS solutions at *upper layers*, such as the *application and middleware layers*, integrated with MAC/network layers to assist in adapting and providing *end-to-end proportional delay differentiation guarantees* in IEEE 802.11 wireless networks [28, 29]. In Section 4.3, we present QoS solutions with integrated *OS, middleware and network layers* to assist in adapting and providing *end-to-end statistical delay guarantees* in IEEE 802.11 wireless LANs [30]. Section 4.4 provides a summary to this chapter.

4.2 DELAY CONTROL WITH UPPER LAYERS ADAPTATION

Achieving QoS guarantees with appropriate QoS and resource services is challenging. As we already discussed in Chapter 3, continuous estimation of available resources, calculation of QoS and resource status at times t, and enforcement of admitted resource usage via techniques such as *cross-layering*, *resource coordination*, and *dynamic resource management* at network layers (transport, IP, and MAC layers) is a must.

In this section, we take into account not only the network layers, but also the upper layers such as middleware and application layers when provisioning end-to-end delay. We present an *upper layer adaptation framework*, integrated via cross-layering with network layers. We will first discuss the overall cross-layer framework in Section 4.2.1, followed by the network/MAC layer *Differentiation Scheduler* design in Section 4.2.2, description of the upper layer *Adaptors* design in Section 4.2.3, and practical issues when relying on upper layer adaptation in Section 4.2.4. Throughout the Section 4.2, we will rely on definitions and notations in Table 4.1.

TABLE 4.1: Notations and definitions for Section 4.2.

NOTATIONS	DEFINITIONS
K	Number of network service classes
$\overline{d}_i(t, t+\tau)$	Average delay for class i over time interval $[t, t+\tau]$
$\boldsymbol{D} = \{D_1, \ldots, D_r\}$ with $D_1 < D_2 < \ldots < D_r$	Set of multiple delay levels

(*continued*)

TABLE 4.1: *(continued)*

NOTATIONS	DEFINITIONS
$\mathrm{d}(t)$	Measured round-trip delay for packet p at time t
$d_{\mathrm{req}} \in [d_{\mathrm{min}}, d_{\mathrm{max}}]$; $\mathrm{ref}_{\mathrm{delay}} := d_{\mathrm{req}}$	Required (desired) delay; d_{req} becomes the reference value for priority adaptor
$d_{\mathrm{min}} = d_{\mathrm{req}} - {}^{j_{\mathrm{req}}}\!/_{2}$	Lower bound of the desired delay
$d_{\mathrm{max}} = d_{\mathrm{req}} + {}^{j_{\mathrm{req}}}\!/_{2}$	Upper bound of the desired delay
$e(t) = \mathrm{d}(t) - d_{\mathrm{req}}$	Error between measured delay and desired delay
j_{req}	Required jitter
M	Number of end-to-end delay measurements to average over
δ_i	Service differentiation parameter for class i
$n \in \mathrm{N}$	Set N of wireless nodes n
$(t, t + \tau)$	Time interval with τ as the monitoring time scale
$\widehat{w}_p(t,i)$	Normalized queue waiting time at time t
$p \in P$	Packet p in set P of all backlogged packets
$q \in Q$ with $\mathbf{Q}$ priorities	Set Q of priorities 1 with $\lvert Q \rvert = \mathbf{Q}$
k_p	This parameter represents the *proportional gain* in the PI (Proportional-Integral) controller
k_i	This parameter represents the *integral gain* in the PI controller

4.2.1 Cross-Layer QoS Framework

The upper layer adaptation relies on the cross-layer protocol stack that includes five major components as shown in Figure 4.1: the MAC/network cross-layered *differentiation scheduler*, based on distributed Waiting Time Priority (WTP) packet scheduling policy, the middleware-based components such as the *delay monitor*, *classifier*, *priority adaptor*, and the application-specific *requirement adaptor*.

Proportional delay differentiation scheduler. This scheduler implements the *proportional delay QoS service differentiation model* which was first time introduced as a per-hop behavior (PHB) for Diffserv (Differentiated Services) in the context of wireline Internet networks [7]. The basic idea of proportional differentiation is that even though the actual quality level (e.g., delay) of each service class will vary with traffic loads, the quality ratio between classes will remain constant in various-sized timescales. Further, such a quality ratio can be controlled by setting service differentiation parameters, which provide great flexibility in class provisioning. We will discuss details of the proportional delay service differentiation model in Section 4.2.2. Suffice to say at this point that one of the packet scheduling algorithms that can enforce the *proportional delay differentation* in short

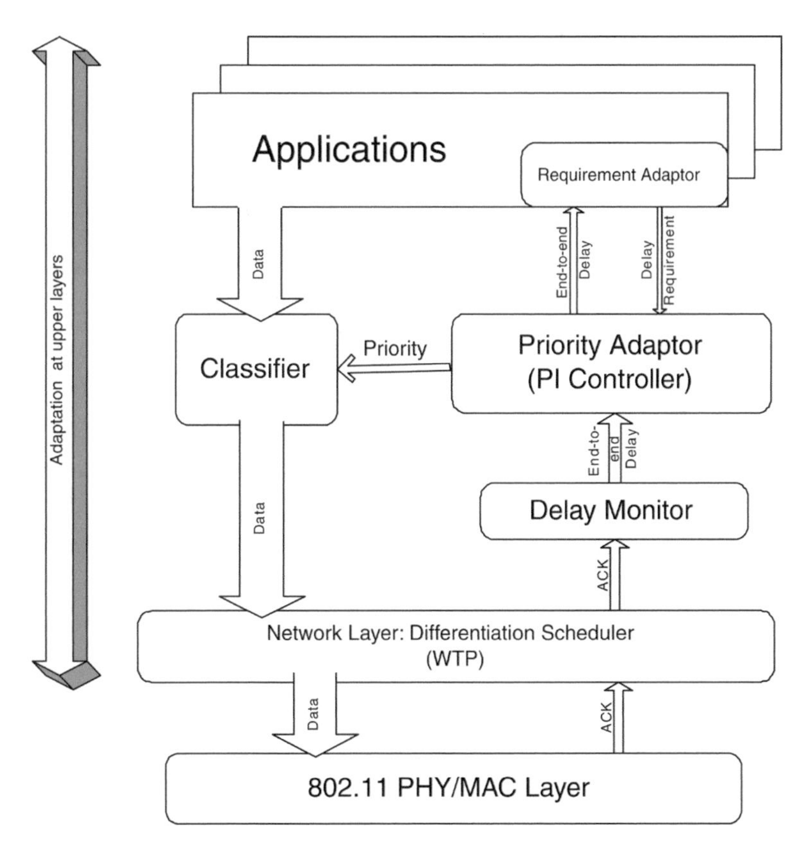

FIGURE 4.1: Cross-layer framework with upper layer adaptation [29].

timescales is the *waiting time priority (WTP)* scheduler [23]. In our design of the differentation scheduler, we present the *distributed cross-layer WTP scheduler* between MAC and network layers for our overall delay management. Details of the scheduler design will be provided in Section 4.2.2.

Delay monitor: The delay monitor resides in the middleware layer, and it measures the average round trip delay incurred to deliver multimedia packets for each application. The measured end-to-end latency contains the delay introduced by traversing the entire protocol stack at the end nodes. Note that we assume wireless reachability and connectivity between two end nodes when they perform real-time multimedia communication. The delay manager is placed in the middleware layer to measure a true end-to-end delay measurement, and it anticipates that lower protocol stack layers ensure *connectivity* and *reachability* even if nodes move. The sender node attaches *timestamps* when sending packets to the destination. When ACK arrives from the destination, the sender retrieves the sending timestamp, compares it with the current timestamp to obtain the round-trip delay d_i for packet p. We take M round-trip delay measurements ($d_1, d_2, \ldots, d_M$) and compute average end-to-end delay $\overline{d}$, $\overline{d} = \frac{1}{2M}\sum_{i=1}^{M} d_i$ to estimate end-to-end delay from the round trip delay measurements. It will be then used in the *Priority Adaptor* to update the service priority appropriately. The priority adaptor will be discussed in Section 4.2.3.

Classifier. This component is also embedded in the middleware layer, and it determines the service classes for sending packets according to their priorities. The goal of the classifier is to *map* **Q** priorities to K network service classes with **Q** $> K$. Note that applications may have access to a large priority space, but as we move down in the protocol stack, the priority space gets smaller for performance reasons. It means, large priority space in the lower network layers introduces service overhead, which impacts *throughput* and *timing* performance of applications and their underlying supporting system. This mapping is very similar to the *priority mapping* between network and MAC layers, which we will discuss in Section 4.2.2. The classifier knows each network service class's parameter, a number obtained by rounding up the application packet priority in a certain range. The application packet priorities will be generated by the *Priority Adaptor*, and the classifier divides the application packet priorities into K ranges $R_1, R_2, \ldots, R_K$. If the priority attached to the application packet by the Priority Adaptor is in the range R_i, then the packet will be assigned to the network service class i with its highest priority within the class i.

Adaptors. We will consider two adaptors within the framework, the *requirement adaptor* and the *priority adaptor*. We only briefly discuss the interrelations between the two adaptors and their impact on the other components of the framework. The details of the adaptors will be discussed in Section 4.2.3. The requirement adaptor gets quantifiable metrics, *end-to-end latency*, and its variance (*jitter*) from the application. For example, in telephony, one-way delay requirement ranges from 25 to 400 ms, and jitter is set to be 20 ms. The requirement adaptor takes the range of requirements, divides the requirements into quality levels, and during the application runtime, it selects the desired delay quality level, and passes it as the reference value $\mathrm{ref}_{\mathrm{delay}}$ to the priority adaptor. The priority adaptor takes the *delay reference value* $\mathrm{ref}_{\mathrm{delay}}$ from the Requirement Adaptor, the *measured QoS delay value* $d_i(t)$ from the delay monitor, and calculates the desired packet priority q which will be then given to the Classifier. (Note that the priority calculation details will be presented in Section 4.2.3.) The Classifier maps the calculated desired application packet priority q into the network packet priority of the network service class to be used by the network WTP differentiation scheduler.

4.2.2 Proportional Delay Differentiation Scheduler

As discussed above, the proportional QoS service differentiation model was first introduced as a PHB (Per-Hop-Behavior) for DiffServ in the context of wireline networks [7]. The proportional differentiation relation is defined as a quality ratio of average delays $\bar{d}_i(t,t+\tau)$ and $\bar{d}_j(t,t+\tau)$ from two different network service classes i and j, as shown in (Equation 4.1.) [7]. The relation indicates that even though the actual quality (e.g., delay) of each service class might vary with traffic load, the quality ratio between classes will remain constant in various-sized timescales.

In this section, we consider the proportional delay differentiation model in the domain of wireless single hop networks such as IEEE 802.11. In IEEE 802.11 network, which consists of a set of nodes N, the traffic is sent among different pairs of nodes and is divided into K network service classes. The proportional differentiation relation in Equation 4.1 holds for traffic between any pairs of nodes. It is important to stress that in wireline networks, the proportional differentiation model only applies for traffic originated from the same router in wireline networks, it applies from different source nodes.

$$\frac{\bar{\mathrm{d}}_i(t,t+\tau)}{\bar{\mathrm{d}}_j(t,t+\tau)}=\frac{\delta_j}{\delta_i},\forall i\neq j\wedge i,j\in\{1,2,...,K\} \qquad (4.1)$$

where δ_i is the service differentiation parameter for class i, and $\bar{\mathrm{d}}_i(t,t+\tau)$ is the average delay for class i, $(i = 1,2,\ldots, K)$ in the time interval $(t, t+\tau)$ with τ as the monitoring time scale.

One of the packet scheduling algorithms that can enforce the proportional delay differentiation in short timescales is the *waiting time priority (WTP) scheduler* [23]. In this algorithm, a network packet in the network queue is assigned a *weight* which increases proportionally to the packet's *waiting time*. Network service classes with higher differentiation parameter have larger *weight-increase* factors. The packet with the largest weight value is served first in *nonpreemptive* order. Formally, let $w_p(t)$ be the waiting time of a network packet p of network class i at time t, we define its *normalized waiting time* $\tilde{w}_p(t,i)$ at time t as follows:

$$\tilde{w}_p(t,i)=w_p(t)\cdot\delta_i \qquad (4.2)$$

The normalized waiting time is then used as the weight for differentiation scheduling. The packet with the largest weight is always selected by the WTP for transmission. Formally, at time t, it will transmit the network packet p which satisfies

$$p = \arg\max_{p \in P} \widehat{w}_p(t,i), \tag{4.3}$$

where P is the set of all backlogged packets. It has been shown that WTP scheduling algorithm approximates the proportional delay differentiation model in the wireline network under heavy traffic condition [23]. However, we need a different approach for the wireless domain. The original WTP in wireline networks is a *centralized scheduling scheme*, and it needs to know the waiting times of all packets before deciding which one to transmit. This is trivial in a wireline network, where all packets waiting to be scheduled originate from the same node. However, in the wireless single hop network, packets that need to be scheduled originate from different source nodes. To acquire information of all packets inside the network, a node has to constantly receive reports from all other nodes, which is prohibitively expensive.

Hence, we will briefly discuss a possible solution, the *distributed and cross-layer scheduling framework (DWTP)* [2], to achieve proportional delay diferentiation in wireless single hop networks with low overhead. The solution takes the *cross-layer approach* between the network and MAC layers, as well as *two-tier distributed WTP scheduling* approach, where at the first tier, *intranode scheduling* at node n selects a packet p_n^* which satisfies

$$p_n^* = \arg\max_{p \in P_n} \widehat{w}_p(t,i), \tag{4.4}$$

where P_n is the set of all backlogged packets at node n. At the second tier, internode scheduling selects packet p^* among $p_n^*, n \in N$, which satisfies

$$p^* = \arg\max_{p_n^*, n \in N} \widehat{w}_{p_n^*}(t,i). \tag{4.5}$$

Such a two-tier intra- and internode scheduling algorithm can fit well the wireless single hop environment.

Cross-layer scheduling architecture. The cross-layer architecture to deploy DWTP consists of the *packet scheduling function* at the network layer and the *distributed coordination function* at the MAC layer where both functions are coordinated via *cross-layer signal.* Note that this approach means some augmentation of the IEEE 802.11 MAC layer to assist in delivering proportional delay differentiation guarantees. Figure 4.2. shows the cross-layer architecture, where at each node, WTP is used for intranode packet scheduling.

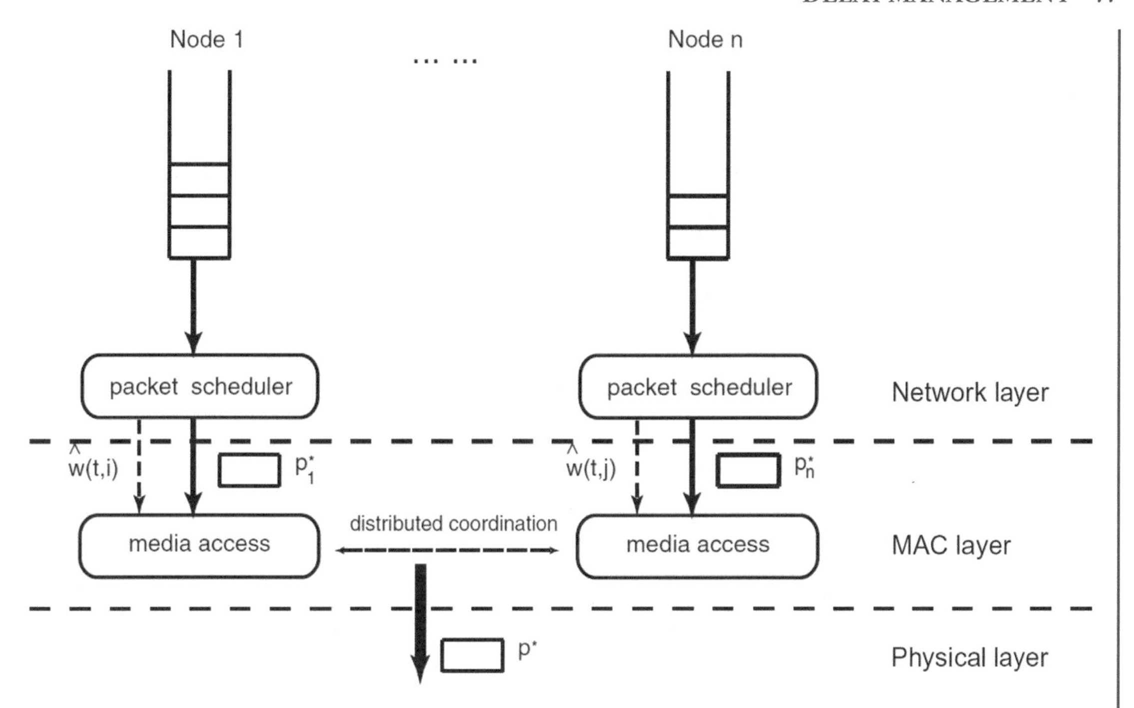

FIGURE 4.2: Two-tier distributed WTP scheduling framework [28].

The WTP at each node selects the packets with the longest normalized waiting time and sends it to the MAC layer. At the same time, it notifies the MAC layer about the normalized waiting time of this packet. At the MAC layer, the packet with the longest normalized waiting time will be transmitted via distributed coordination function among the nodes. One way to achieve such transmission is to use *priority-enabled MAC* [24, 25, 26, 27], where packets with higher priorities will be transmitted before packets with lower priorities. However, for the purpose of proportional delay differentiation, the following two issues have to be solved before DWTP can be properly deployed.

First, to utilize the *priority-enabled MAC*, packets with different normalized waiting times at the network layer need to be assigned with *different MAC priorities*, such that the packets with longer normalized waiting times will be transmitted before packets with shorter normalized waiting times. This presents a problem of *priority mapping*. As a priority-enabled MAC only supports a *small number of priorities*, there may exist network packets with different normalized waiting times that are assigned with the same MAC priority. In this case, the packet with the largest waiting time may be assigned with the same priority as packets with smaller waiting times and may be transmitted at a later time than those packets, causing a problem of *priority inversion*.

Second, at the network layer, once a packet is selected and delivered to MAC layer, it will be served in a *nonpreemptive manner* and cannot be sent back to the network layer. This may cause another type of *priority inversion* problem. For instance, suppose at node n, packet p_n' has the largest normalized waiting time $p_n' = \arg\max_{p \in P_n} \widehat{w}_p(t_{mac}, i)$ at time t_{mac}, and is selected to be delivered to the MAC layer. Owing to the distributed coordination at the MAC layer, packet p_n' may not be transmitted immediately. Instead, it may actually be transmitted at time t_{tx}. Yet at time t_{tx}, the normalized waiting time of the packet p_n', $\widehat{w}_{p'}(t_{tx}, i)$ may no longer be the largest at node n. It is possible that another packet p_n^* in the network layer queue now has the largest waiting time, i.e., $p_n^* = \arg\max_{p \in P_n} \widehat{w}_p(t_{tx}, i)$, and therefore, it should be transmitted first according to the WTP algorithm. We will address both issues below.

Two-tier distributed WTP scheduling. The two-tier distributed WTP scheduling framework includes the internode scheduling and the intranode scheduling. As discussed above, the internode scheduling selects packet p* among $p_n^*, n \in N$, which satisfies Equation 4.5. The intranode scheduling at node n selects a packet p_n^*, which satisfies Equation 4.4. The internode scheduling will address the priority mapping problem, and the intranode scheduling will address the priority inversion problem, as discussed above.

The *priority mapping problem* means that we have more normalized waiting times of network packets (and their priorities) than MAC priority levels; hence, different normalized waiting times of network packets will be mapped into the same MAC priority. Let us consider a priority-enabled MAC protocol with **Q** priorities. For two packets p and p' with priorities q and q', if $q > q'$, then packet p will be transmitted before packet p'. If $q = q'$, then the packets p and p' will be transmitted in a random order. Under such a MAC model, transmitting the packet with the largest normalized waiting time $\widehat{w}_{p_n^*}(t, i)$ requires the highest priority to be assigned to this packet. Formally, we have the following *ideal mapping condition* under which the *priority-enabled distributed coordination* can precisely implement the internode scheduling defined in Equation 4.5.

Ideal mapping condition. Let $\widehat{W}(t)$ be the set of normalized waiting times of network packets coming to MAC layer at time t, and $Q = \{1, 2, \ldots \boldsymbol{Q}\}$ be the set of MAC priorities. A map function $\phi : \widehat{W}(t) \rightarrow Q$ is an ideal mapping, if it satisfies the following: for $\widehat{w}^* = \max_{\widehat{w} \in \widehat{W}(t)} \widehat{w}$, the largest normalized waiting time at time t, it is true: $\phi(\widehat{w}^*) = q$ if and only if $\phi(\widehat{w}') = q'$ and $q' < q$ for $\forall \widehat{w}' \in \widehat{W}(t), \widehat{w}' \neq \widehat{w}^*$. Such an ideal mapping function is hard to implement because the set of normalized waiting times $\widehat{W}(t)$ is changing over time t. Hence, tracking the element in this set is very difficult in a distributed environment. We consider a more practical method, called fixed priority mapping. For example, Figure 4.3 illustrates that in fixed priority mapping, there are $\boldsymbol{Q} - 1$ cut-off points, $\widehat{w}_q^*$, $q = 1, \ldots, \boldsymbol{Q} - 1$.:

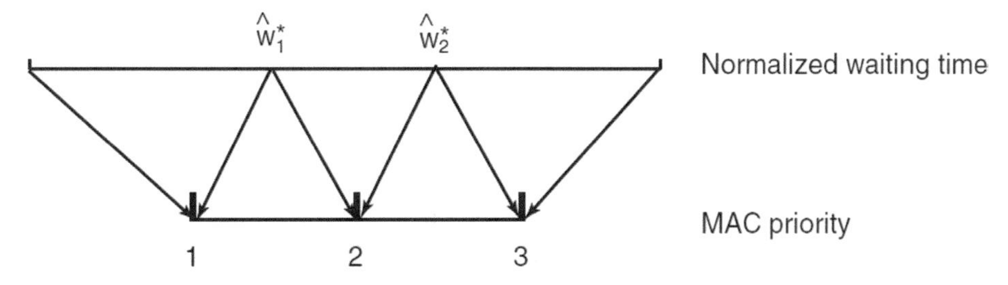

FIGURE 4.3: Fixed priority mapping [28].

It means, if a normalized waiting time $\hat{w} \in [\hat{w}_q^*, \hat{w}_{q+1}^*)$ with $Q-1$ cut-off points and $q = 1, \ldots, Q-1$, then $\hat{w}$ maps to priority q (note that $\hat{w}_0^* = 0; \hat{w}_Q^* = \infty$). This mapping function is simple to implement. However, the selection of the cut-off points could have large impact on the delay differentiation result.

Figure 4.4 shows a simple simulation of the impact of different cut-off point settings. The simulation uses two flows of two network service classes with $\delta_1 = 1$, $\delta_2 = 2$ between two different pairs of nodes. The results in Figure 4.4 show that with proper cut-off point setting, e.g., $\hat{w}_1^* = 0.3$, the fixed priority mapping can achieve proper internode scheduling for proportional delay differentation. However, if $\hat{w}_1^*$ is set to 0.2 or 0.5, proper internode sceduling cannot be achieved. The reason behind this result can be explained in Figure 4.5.

Figure 4.5a illustrates the centralized WTP scheduling, where a packet from class j is transmitted at time t as its normalized waiting $(t - \tau)\delta_j$ is larger than that of a packet $t\delta_j$. Figure 4.5b shows the scenario of distributed WTP. If the cut-off point is properly set, the packet from class j will be transmitted first, maintaining the priority relationship. However, if $\hat{w}_1^*$ is set too large (e.g., $\hat{w}^{*'}$) or too small (e.g., $\hat{w}^{*''}$), the packets are assigned to the same priority, which may lead to *priority inversion* as we described earlier.

To remedy the priority inversion for the distributed WTP scheduling, we introduce W_{max} to be the random variable that represents the largest normalized waiting time of all packets at the MAC layer, and W_{min} be the random variable that represents the rest of normalized waiting times. If we would know the distribution functions of W_{max} and W_{min}, we could achieve optimal cut-off parameter settings, but this is not feasible in a practical network enviroment. Hence, we adjust the cut-off parameter w^* so that it approximates the average of the expectation values of W_{max} and W_{min}. Specifically, let $t_1, t_2, \ldots, t_m, \ldots$ be a sequence of time instances at which $\hat{w}^*$ is adjusted. Let $E(W_{max})([t_{i-1}, t_i])$ and $E(W_{min})([t_{i-1}, t_i])$ be the expected value of the normalized waiting times of

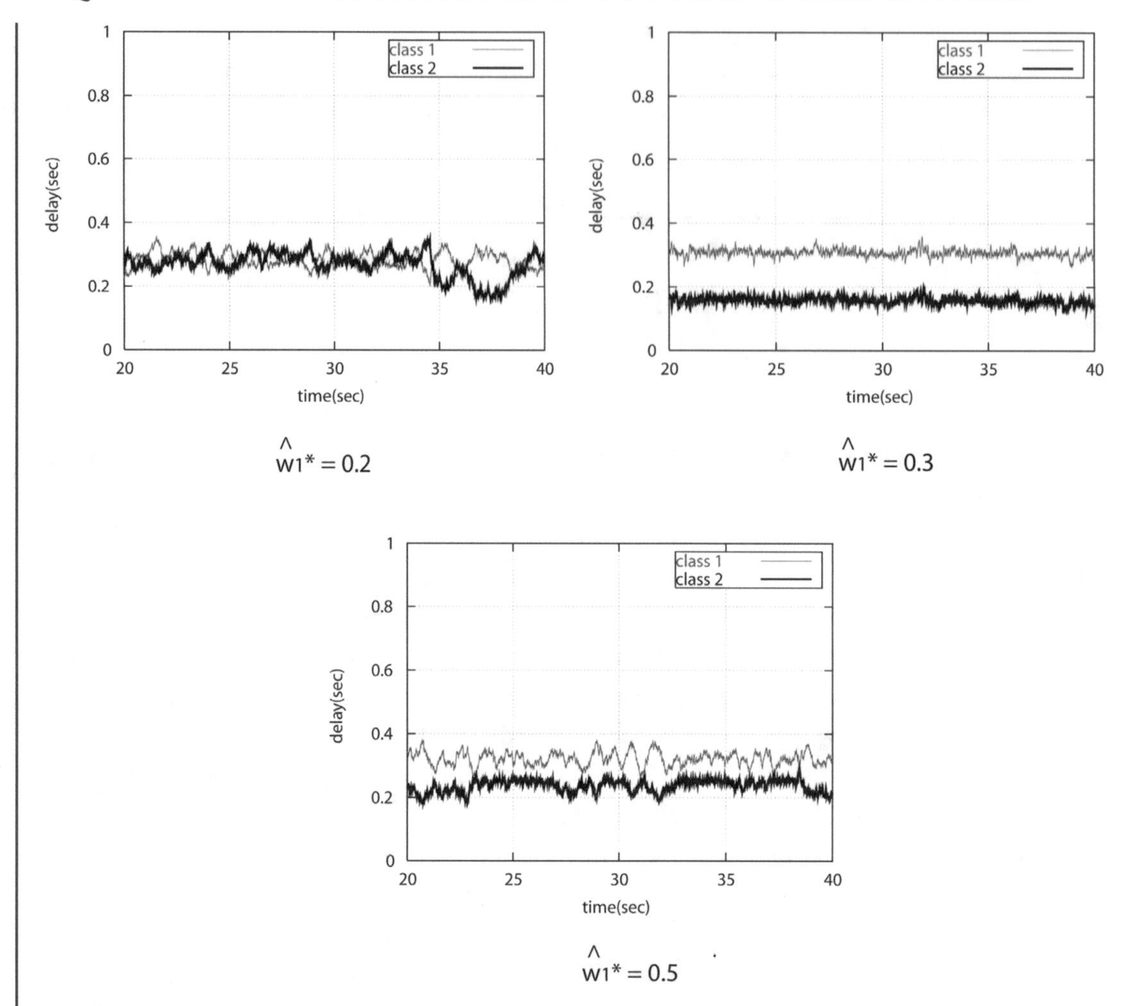

FIGURE 4.4: Impact of different cut-off point settings for delay differentiation [28].

high priority packets and low priority packets, respectively, during the last observation time window $[t_{i-1}, t_i]$. Then, in the next time window $[t_i, t_{i+1}]$, the cut-off parameter ϖ^* is set to

$$\varpi^* = (E(W_{\max})([t_{i-1}, t_i]) + E(W_{\min})([t_{i-1}, t_i]))/2 \tag{4.6}$$

The priority inversion may be also caused by layering because once a packet is selected and delivered to the MAC layer, it will be served in a nonpreemptive manner. To address this problem, the *intranode packet scheduler* adopts a *predicted waiting time priority scheduling algorithm*, which works as follows.

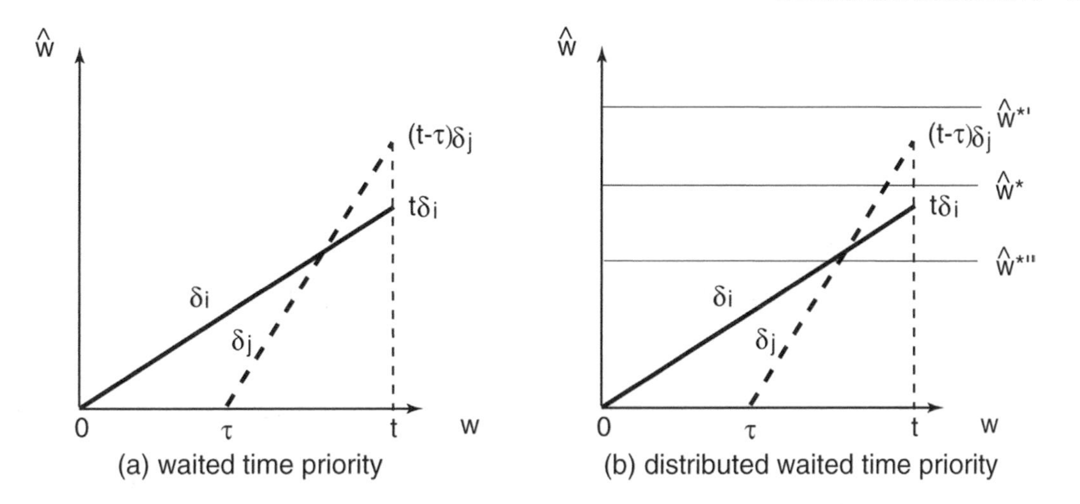

FIGURE 4.5: Priority inversion with improper cutting off parameter setting [28].

At time t_{mac}, when a packet is to be selected, instead of choosing the packet with the largest normalized waiting time $p_n' = \arg\max_{p \in P_n} \widehat{w}_p(t_{mac}, i)$ at current time t_{mac}, it selects the packets with the largest normalized time at the *predicted transmission time* $\overline{t}_{tx}$, $p_n^* = \arg\max_{p \in P_n} \widehat{w}_p(\overline{t}_{tx}, i)$. The predicted transmission time is based on the measurement of history transmission delay in the last time window.

4.2.3 Adaptors Design

The upper layer adaptation framework, to achieve proportional delay quality guarantees, consists of two adaptors, the requirement adaptor and the priority adaptor. These adaptors will be described relying on the basic concepts of the *control theory*. The basic concept in control theory is the "control loop," shown in Figure 4.6a that exists between the controlled System, the Monitor that monitors the status of the System, and the Controller that adapts the system based on the feedback (measured error). The Controller receives the *measured error*, which is the difference between the System's desired quality level and the *measured quality level* (measured output), coming from the Monitor.

In our wireless control loop scenario, shown in Figure 4.6b, the *System (G)* is the *end-to-end wireless transmission system* of multimedia data, including the *multimedia application* (e.g., VoIP), *application packet classifier*, and the network/MAC layer *differentiation scheduler* as shown in Figure 4.1. In the middleware layer, we have the *delay monitor* at the sending side of the transmission system that serves as the Monitor (M). The measured output in the wireless delay scenario is the observed delay d at step t (where steps are time measurement intervals). The wireless control loop is completed by the Controller (C), which resides at the sending side of the transmission system. The Controller receives error $e(t)$ which is the difference between the reference delay ref_{delay} and $d(t)$.

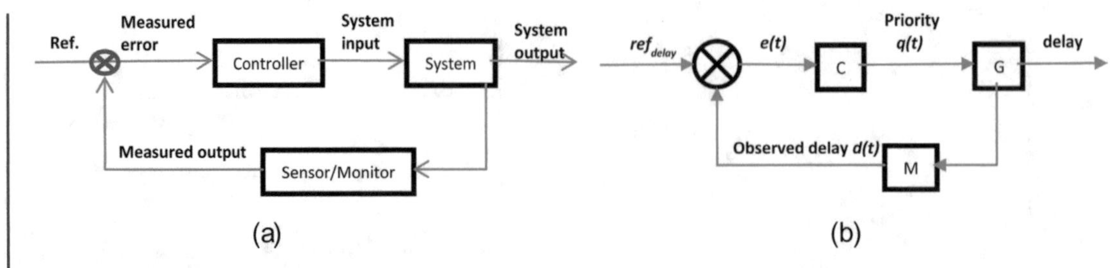

FIGURE 4.6: (a) Generic control loop to deal with dynamic systems and (b) control loop for our wireless system delay control.

Note that the *Requirement Adaptor* in our wireless scenario represents the component that provides the delay reference values to the Controller, and the *Priority Adaptor* is the Controller component that calculates the error $e(t)$ and the application priority q for packets. (Note that in the following text, we will use the terms Controller and Priority Adaptor interchangeably.) In Sections 4.2.1 and 4.2.2, we discussed the System (Differentiation Scheduler and Classifier) and Delay Monitor components. Below, we will discuss the details of the Adaptors.

Requirement adaptor design. This adaptor takes the delay QoS requirement from the multimedia application, given by the application/user expectations of the *end-to-end delay*. Usually, the expectation is the best quality level of the multimedia delivery that can be achieved under light network load conditions. However, under varying network load conditions, when the best quality is not achieved, users may prefer to tolerate quality degradation rather than drop the application.

We model delay requirements for these types of adaptive applications as follows: An application defines multiple quality levels $\boldsymbol{D} = \{D_1, D_2, ..D_r\}$ in the form of application acceptable delay expectations $D_1 < D_2 < D_3 < \ldots < D_r$. $\boldsymbol{D}$ represents the set of multiple delay levels, d_{req} denotes the required end-to-end delay and j_{req} denotes the required jitter. The *application layer QoS adaptation* in the requirement adaptor is then in charge to select a delay expectation d_{req} from the set $\boldsymbol{D}$ according to the network load condition. Note that different delay expectations in the set $\boldsymbol{D}$ result in different audio or video qualities.

Since deterministic end-to-end QoS guarantees over contention-based (e.g., IEEE 802.11) networks are impossible, we will use statistical and proportional criteria as discussed in Chapter 1 for the delay requirement selection.

With d_{req} and j_{req} given, we define d_{min} and d_{max} as the lower and upper bounds of the desired delay d_{req}. We desire that the end-to-end delay d_{req} falls into the range $[d_{\text{min}}, d_{\text{max}}]$ with $d_{\min} = d_{\text{req}} - \frac{j_{\text{req}}}{2}$ and $d_{\max} = d_{\text{req}} + \frac{j_{\text{req}}}{2}$. $\Pr(d \in [d_{\min}, d_{\max}])$ denotes the probability that delay QoS requirement d within the range $[d_{\text{min}}, d_{\text{max}}]$ is satisfied. We recommend that

$$d_{\text{req}} = \min\{d \in D \mid \Pr(d \in [d_{\min}, d_{\max}]) > 95\%\}. \tag{4.7}$$

Hence, the requirement adaptor does the following steps:

1. It gets the information about the network load conditions (via measured delay output $d(t)$ from the Delay Monitor);
2. It selects the highest possible requirement d_{req} from the requirement set $\boldsymbol{D}$ according to Equation 4.7. which is close to $d(t)$;
3. It passes d_{req} as the delay reference value $\text{ref}_{\text{delay}}$ to the priority adaptor.

In summary, the requirement adaptor's main function is to maintain the discrete set $\boldsymbol{D}$ of desired delay quality levels, adjust the requested delay quality level d_{req} according to the observed network load and give the priority adaptor realistic reference $\text{ref}_{\text{delay}}$ for further adaptation of the overall System.

Priority adaptor design. To design the priority adaptor, we will need to go through two steps: (1) *System Model Identification* to identify the relation between the end-to-end delay $d(t)$ and priorities $q(t)$ in the wireless system; and (2) *Design of the PI Controller* to calculate packet priorities $q(t)$ at each measurement step t.

System model identification. To obtain the model of System G precisely is difficult due to the complexity of the wireless network systems. Hence, we treat the wireless network system as a black box and then infer the model from externally observable metrics. The process to infer the open-loop system model is called *model identification.* The model identification metrics are specific to a particular observed system. In our discussion, we will use the IEEE 802.11 wireless testbed, shown in Figure 4.8b, to gather data for the model identification. Note that in the System model identification, we look at the priority as the System input and the end-to-end delay as the *System* output. On the other hand, in the priority adaptor design, the priority is the output of the *Controller*, and the end-to-end delay is the input (reference delay $\text{ref}_{\text{delay}}$ and measured delay $d(t)$). In order to develop an *adaptive priority adjustment algorithm* to control the end-to-end delay, we need first to understand *how priority affects the end-to-end delay under a certain traffic load* and this is done by system model identification techniques [31]. The standard system identification techniques come from control engineering and use statistical methods to build mathematical models of dynamic systems from measured data.

We use *difference equation* with unknown parameters as the dynamic System model between the input (priority) and the output (end-to-end delay). Such a model estimates the mathematical relation between the input and output of the System. We then use a pseudo-random digital white noise generator to stimulate the System by assigning random priorities to multimedia packets and observe the end-to-end delay during a certain time measurement period. We choose a sampling interval of 0.5 s. Hence, we measure a *(priority, delay) pair* every 0.5 s. With the data we obtain in the experiments, we can then apply the *autoregressive (AR) predictive model* to get the mathematical relationship from the priority to the end-to-end delay. Note that the first-order AR model in Equation 4.8. provides reasonably good prediction for end-to-end delays with different priorities, but other predictive models can be utilized in the System model identification. We use

$$d(t+1) = b_0 q(t) + a_1 \, d(t), \tag{4.8}$$

where $d(t)$ and $q(t)$ are the end-to-end delay and priority at time t, respectively. The relation between $d(t)$ and $q(t)$ in Equation 4.8 is independent of the number of nodes on the route as long as the route between two end-nodes is fixed. b_0 and a_1 are determined in the model identification from the measured (priority,delay) pairs in a considered wireless network system. For example, if we consider 50 input–output pairs of $(q(t), d(t))$ from the testbed in Figure 4.8b, we can identify the wireless system model, i.e., the coeficients b_0 and a_1 as $b_0 = 0.02425$ and $a_1 = 0.2514$. Figure 4.7 shows the comparison between the AR model of the predicted delay and the measured output delay.

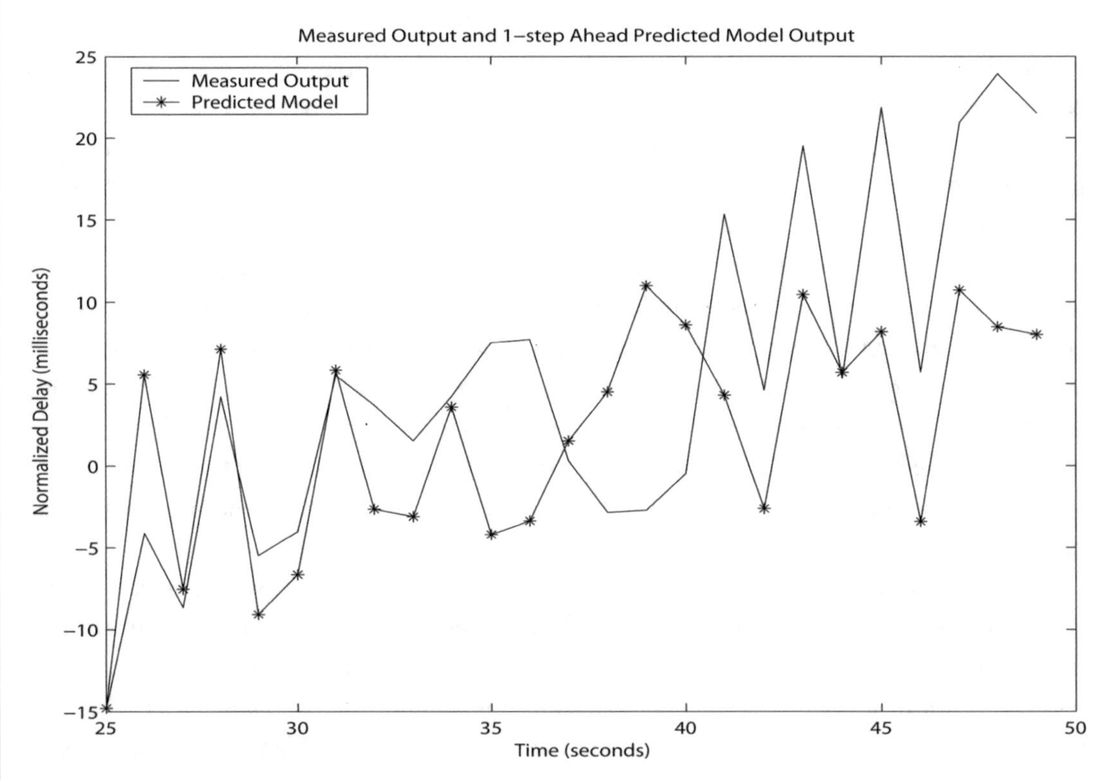

FIGURE 4.7: Comparison of measured output delay vs delay model prediction (Equation 4.8) [29].

Equation 4.8. then resides within the System G, where the System calculates output $d(t+1)$ based on the adjusted priority $q(t)$ coming from the priority adaptor and delay $d(t)$ observed in the previous time measurement interval t by the delay monitor.

Design of PI controller for priority adaptor. The priority adaptor's function is to adjust the priority q, based on observed delay $d(t)$ and reference value $\text{ref}_{\text{delay}}$. We will design this function as the *PI (Proportional and Integral) controller*. The PI controller is one of the standard control techniques to adjust dynamic systems, and it allows us the determination of the new priority $q(t+1)$ to which we need to adjust the next application packet priority. The Priority Adaptor's PI control function consists of two terms, the *proportional* and the *integral* terms.

- The *proportional term* makes a change to the System output proportional to the current measurement error value $e(t)$. In our wireless scenario $e(t) = d(t) - \text{ref}_{\text{delay}}$, i.e., we want the error at time t between the measured end-to-end delay $d(t)$ and the suggested reference delay $\text{ref}_{\text{delay}}$. The proportional response can be adjusted by multiplying the error by a constant k_p, called *proportional gain* (i.e., output value from the proportional term is $k_p e(t)$). A high proportional gain results in a large change in the System output for a given change in the error. If the proportional gain is too high, the System becomes unstable. In contrast, a small gain results in a small output response to a large input error and a less responsive or less sensitive Controller. In our setting, the *proportional* term of the priority adaptor has the effect of reducing the rise of the end-to-end delay time and decreasing the steady-state error.
- The contribution from the *integral term* is proportional to both the magnitude of the measurement error and the duration of the error. The integral term in a PI controller is the sum of the instantaneous errors over time $\sum_{\tau=0}^{t} e(\tau)$ and gives the accumulated offset that should have been corrected previously. The accumulated error is then multiplied by the integral gain k_i and added to the Controller output (i.e., the output value from the integral term is $k_i \sum_{\tau=0}^{t} e(\tau)$). The integral term accelerates the movement of the process toward setpoint and eliminates the residual steady-state error that occurs with a pure proportional controller. However, since the integral term responds to accumulated errors from the past, it can cause the present value to overshoot the setpoint value, so one has to be careful. In our priority adaptor design, the *integral* term has the effect of eliminating the *steady-state error*, but it may make the transient response worse.
- Some Controllers also use a *derivate term* in their controller design (so called PID controllers). The derivate of the process error is calculated by determining the slope of the error over time and multiplying this rate of change by the *derivative gain* k_d. The derivative term

slows the rate of change of the Controller output and has the effect of increasing stability of the system, reducing the overshoot, and improving the transient response. However, if the system noise is large, the derivate controller will decrease the stability of the system. In the wireless system, both the system noise and the measurement noise are large. The *system noise* occurs due to the random workload in the ad hoc network and the nature of randomized algorithms in MAC layer protocols. The *measurement noise* comes from the measured data we get when monitoring round-trip delay $d(t)$. Owing to the large noise in the system, the derivative controller will introduce the undesired oscillation to the system; hence, we will consider only the proportional and integral terms in the priority adaptor.

Priority calculation. As discussed above, the priority adaptor is part of the major control loop, shown in Figure 4.1, which is a concrete example of the generic control loop in Figure 4.6. The priority adaptor runs the PI controller to determine the application packet priority q as follows:

1. It takes the observed delay $d(t)$ and the requested reference value $\text{ref}_{\text{delay}} = d_{\text{req}}$ at time t and computes the measured error $e(t) = d(t) - d_{\text{req}}$. If $e(t)$ can be kept close to zero, we achieve steady state of the System, and the end-to-end delay of the System will stay close to the desired delay quality level.
2. It calculates the *priority control equation* using the proportional and integral terms the PI controller as follows:

We calculate $q(t+1) = k_p e(t) + k_i \sum_{\tau=0}^{t} e(\tau)$ (4.9)

We simplify Equation 4.9 by considering $q(t) = k_p e(t-1) + k_i \sum_{\tau=0}^{t-1} e(\tau)$ (4.10)

We subtract Equation 4.10 from Equation 4.9, and we get the *priority control equation*

$$q(t+1) = q(t) + (k_p + k_i)e(t) - k_p e(t-1) \tag{4.11}$$

The parameters k_p and k_i represent the proportional and integral gains, related to the errors, and are *wireless network specific*, since they depend on the System G (i.e., on b_0 and a_1 coefficients in Equation 4.8., which are obtained via probing measurements of (priority, delay) pairs). The process, how to determine parameters k_p and k_i relies on bringing Equation 4.8 together with Equation 4.11 in an iterative way. (Note: This process to solve Equation 4.8 and Equation 4.11 can be complex in the time domain t. To make the calculations of k_p and k_i simpler, mathematics offers us transformation of Equation 4.8 (system model prediction) and Equation 4.11 (priority control equation) from their discrete time-domain values $d(t)$ and $q(t)$ to their frequence-domain values. The transformation is called *z-transform*. The reason for this transformation is *simplicity*. The detailed calculation of k_p and k_i via z-transform details are described in Reference [29].)

4.2.4 Practical Issues

There are several observations that one needs to take into account when considering this upper layer adaptation framework.

First, the upper layer adaptation design relies on generic components such as classifier and the supporting cross-layer differentiation scheduler. However, the upper layer adaptation components such as the *requirement and priority adaptors* are application-specific and can exist without QoS-specific approaches in network and MAC layers such as differention schedulers and priority-based services. But without any underlying QoS-aware mechanisms, the delay variations increase. Especially, under heavy network load conditions, the adaptors may not be able to control delay within acceptable delay bounds.

Second, wireless IEEE 802.11b experiments confirm that if one has only the network *differentiation scheduler (no priority-enabled MAC)* in place and *does not have any middleware adaptation mechanisms* in place, the performance of multimedia flow(s) under UDP background traffic is poor and increases end-to-end delays. We will show this behavior within two testbed topologies, shown in Figure 4.8. The topology scenario in Figure 4.8a shows three wireless IEEE 802.11b nodes, connected in chain via ad hoc mode, where audio-sending application goes from machine 1 through machine 2 to machine 3. The background traffic is being sent from machine 2 to machine 3. The topology scenario in Figure 4.8b shows five wireless 802.11b nodes, connected in a star formation, where audio *flow 1* sends data from machine 1 to machine 3 through machine 5, *flow 2* sends data from machine 2 to machine 4 through machine 5, and background traffic goes along the paths of each *flow 1* and *flow 2*.

Using these testbeds, we show the end-to-end delay performance without any upper layer adaptation mechanisms in Figure 4.9. Figure 4.9a shows the end-to-end delay of a single multimedia flow with UDP background traffic at rate of 100 kbps. The expected/desired end-to-end delay for this flow is 60 ms. Note that the background traffic starts only after the multimedia application sends 170 packets. Hence, we have very low end-to-end delay for the first 1–170 audio packets (audio rate is 256 kbps). After the 170th packet, we start to see linear increase of the audio packets end-to-end delay, going up to 10,000–14,000 ms (10–14 s) end-to-end delay for 400–500th audio packet which is *unacceptable*. Figure 4.9b shows the end-to-end delay of two audio flows 1 and 2 with the expected end-to-end delay 100 ms for flow 1 and 200 ms for flow 2 and UDP background traffic of 20 kbps. The background traffic starts immediately with the multimedia flows. As we can see, the end-to-end delays for both flows increase to unacceptable bounds going from 500–1500

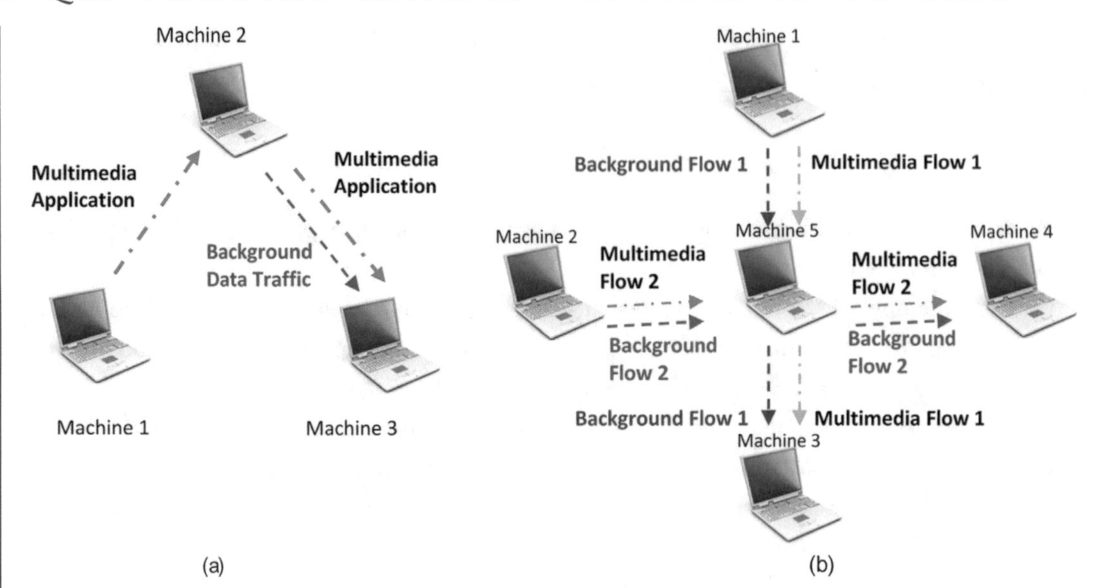

FIGURE 4.8: Testbed scenarios to experiment with end-to-end delays: Scenario 1 with three-node topology and Scenario 2 with five-node topology.

ms delay at the beginning to 3500–4500 ms (Note that the delays are lower in Figure 4.9b than in Figure 4.9a due to much lower background network load condition.). However, one aspect might be interesting to describe in Figure 4.9b. The two flows are maintaining steady distance from each other (but not according to the required delay ratio 2:1 following Equation 4.1). This is attributed to the *network differentiation scheduler*, which is in place (only the middleware adaptation is not available). It starts with the initial measured end-to-end values (600 ms for flow 2 and 1450 ms for flow 1), calculates the delay ratio, and then maintains the delay ratio, since it does not have any knowledge about the proper flows' reference delays from the application.

Third, with the deployment of the middleware adaptation PI controller, *statistical end-to-end delay control* is possible for multimedia flows even if background UDP traffic runs. In Figure 4.10, we show the same audio flows and their end-to-end delay results in Scenario 1 and Scenario 2 testbeds that are controlled by the PI controller $q(t+1) = q(t) + 3.36e(t) - 1.85e(t-1)$. Recall that the PI controller coefficients ($k_p = 3.36$ and $k_i = -1.85$) *are gained* from solving iteratively Equations 4.11 and 4.8. They again depend on the System's coefficients b_0 and a_1, which are gained by probing the wireless network testbed in Scenario 2. We use the same PI controller for both scenarios. We can see that especially in case of two flows in Scenario 2, the flows maintain the proportional 2:1 delay ratio (the original deadlines were 100 ms for *flow 1* and 200 ms for *flow 2*). The reader notices the spikes which come from the network load. When network load turns to be heavier, some

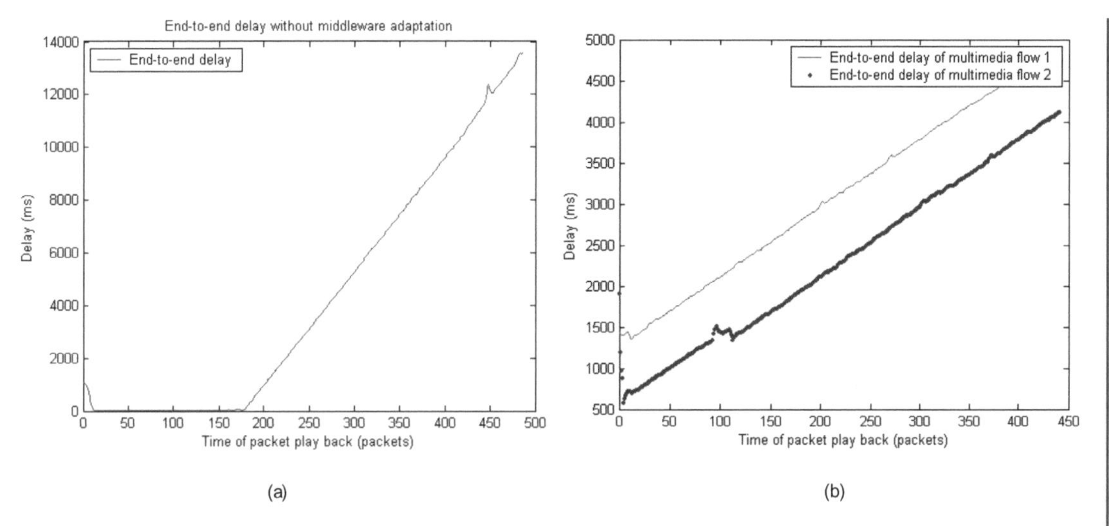

FIGURE 4.9: End-to-end delays without middleware adaptation: (a) delay of single flow under Scenario 1 and (b) delay of two flows under Scenarion 2 [29].

packets suffer very large delays. We adjust priority of packets when we observe large delays, so that the later packets will encounter small delays. Figure 4.10 shows results where each packet keeps its application priority, determined by the priority adaptor, and the classifier does not do any remapping (relabeling) of priorities from application priority to network priority.

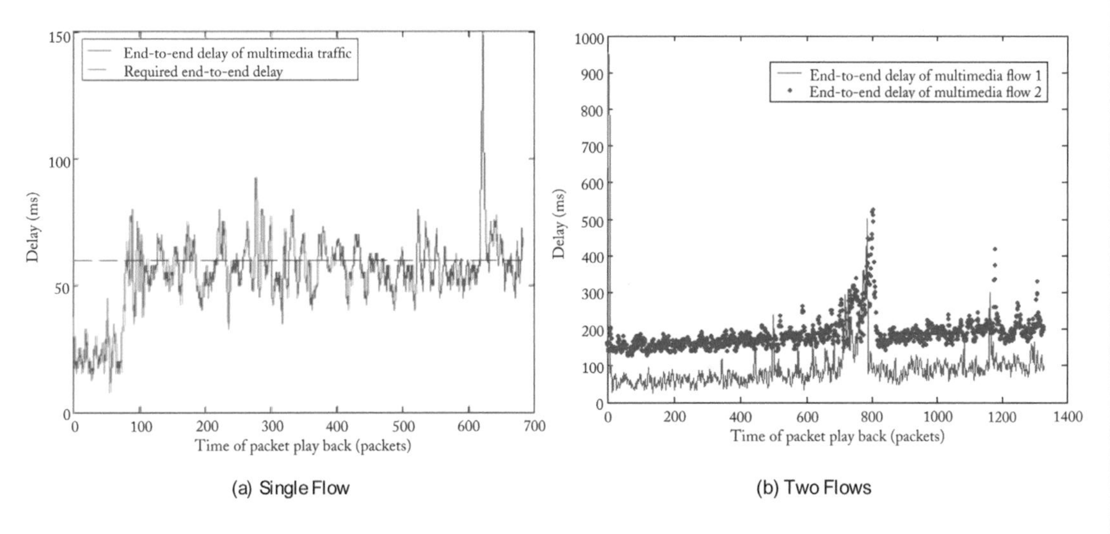

FIGURE 4.10: End-to-end delay control with PI controller adaptation [29].

Fourth, the *classifier's remapping (re-labeling) function* has an important impact on the overall end-to-end delay performance since if a packet belongs to a certain service class, it will be re-labeled with the service class highest priority, i.e., a packet might improve its priority assignment in the classifier component. To show the effect of the Classifier, we will use the Scenario 1 to setup the experiment, and let two multimedia flows send from Machine 1 to Machine 3. The required end-to-end delays of *flow 1* and *flow 2* are 60 and 120 ms, respectively. The background UDP traffic is active during the whole experiment. In this experiment, the middleware Classifier provides mapping from application priorities to 20 network service classes (groups of priorities). Figure 4.11 shows the comparison of multimedia flow control without and with Classifier component.

Comparing the experimental results in Figure 4.11a and 4.11b, one can observe that the controlled end-to-end delay under the Classifier is smaller than the controlled delay without it, i.e., *flow 1* will experience end-to-end delay around 60 ms in Figure 4.11a, and *flow 1* will have end-to-end delay between 40 and 50 ms in Figure 4.11b. To show this decreasing delay effect, the reader should note that the delay *y*-axis in Figure 4.11b shows delay values from 0 to 200 ms indicating finer granularity of delay measurements than in Figure 4.11a, where the delay *y*-axis shows values from 0 to 350 ms. The decreasing delay effect in Figure 4.11b occurs because we select the largest priority in the priority group as the class parameter for each service class.

Fifth, the *cross-layer control framework* (upper layer adaptation plus network layer differentiation scheduler) will yield very good results in proportional service differentiation and statistical end-to-end delay support even without MAC's priority-enabling functions and distributed coordination if the multimedia nodes are all in the same transmission/interference range, or applications share one or more nodes along their routes and the shared nodes can coordinate packet transmission from higher layers.

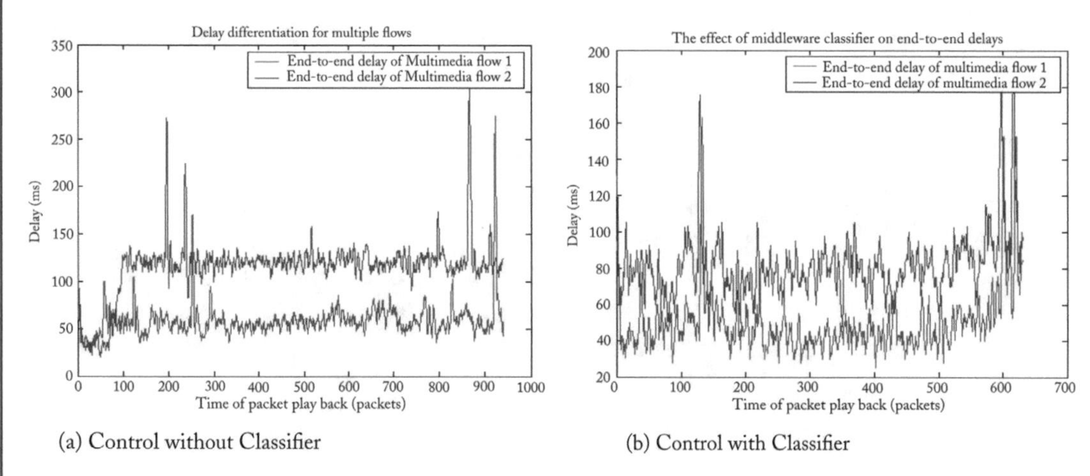

(a) Control without Classifier

(b) Control with Classifier

FIGURE 4.11: Delay control without/with Classifier in adaptation framework [29].

However, if we consider wireless network configurations, where the multimedia flows and background flows do not share common node (do not know about each other), and interfere only implicitly with each other as shown in Figure 4.12, this leads to difficulties in media access control, and problems such as *exposed terminal problem* and *hidden terminal problem*. In these network configurations, the two-tier MAC/network cross-layer scheduling and distributed coordination, as shown in Figure 4.2, must be fully invoked to come close to any QoS solutions. To exemplify the media access control problems, we will briefly explain and show examples of these problems in Figure 4.12.

Exposed terminal problem. Figure 4.12a shows the *flow 1* from S1 to R1 (multimedia flow), where its sending node S1 is in the interference range of the sending node S2 of the *flow 2*, transmitting from S2 to R2 (background flow) because the transmission/interference range of S1 overlaps with the transmission/interference range of S2. Here, if transmission of S1 to R1 is taking place, node S2 is prevented from transmitting to R2 as it concludes via carrier sense that it will interfere with the data transmission by its neighbor S1. However, note that R2 could still receive the transmission of S2 without intereference because it is out of the range of S1. This problem when a node is prevented from sending packets to other nodes due to a neighboring transmitter is called the *exposed terminal problem*.

Hidden terminal problem. This problem, shown in Figure 4.12b, occurs when a node S1 is visible from a wireless access point (*Hub*/*R*), but not from other nodes (e.g., S2) communicating with the said *Hub*. This again leads to difficulties in media access control. Note that *hidden nodes* in a wireless network refer to nodes that are out of range of other nodes or a collection of nodes. For example, let us assume two flows in Figure 4.12b, multimedia *flow 1* from S1 to *Hub* and background *flow 2* from S2 to *Hub*, where S1 and S2 are in transmission range of *Hub*, but not of each other.

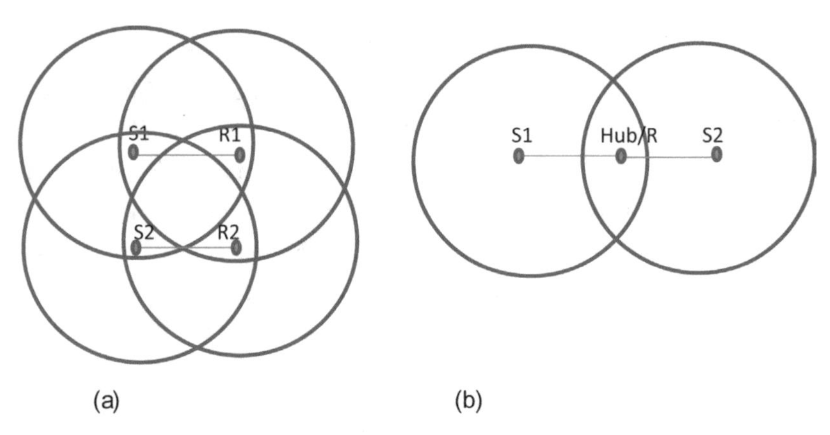

FIGURE 4.12: (a) Exposed terminal problem and (b) hidden terminal problem.

These nodes S1 and S2 are knows as *hidden*. The problem is when nodes S1 and S2 start to send packets simultaneously to the *Hub*. Since node S1 and S2 cannot sense the carrier, *Carrier sense multiple access with collision avoidance (CSMA/CA)* does not work, and collisions occur, scrambling data. To overcome this problem, handshaking (e.g., RTS/CTS approach) or other coordination/scheduling schemes (e.g., scheme from Section 4.2.2) need to be implemented in conjunction with the CSMA/CA scheme. Note that IEEE 802.11 utilizes handshaking to mitigate the hidden terminal problem, which decreases the throughput of the MAC layer. Hence, this problem causes why IEEE 802.11 is suited for bridging last mile for *broadband access only to a very limited extent*. Some newer standards such as WiMAX have now coordination and scheduling mechanisms to assign time slots to individual stations, thus preventing multiple nodes from sending simultaneously and ensuring fairness even in over-subscription scenarios.

4.3 INTEGRATED DYNAMIC SOFT REAL-TIME FRAMEWORK

Critical network infrastructures (e.g., in Power Grid) are starting to deploy general purpose wireless IEEE 802.11 network and computing solutions to decrease the cost of their next generation infrastructure investment. However, these infrastructures require QoS delay control with tighter delay bounds and higher stability in packet delivery than what we saw in Section 4.2 for multimedia applications (e.g., VoIP) over IEEE 802.11 networks. Especially, as systems such as SCADA are coming forward with possible IEEE 802.11 wireless LAN (WLAN) deployments that connect their wireless sensing equipment, e.g., PMUs and IEDs (Intelligent Electronic Devices), with the collection gateway, shown in Figure 4.13, we need to consider deployment of delay-sensitive QoS solutions that consider networks and computation. These QoS solutions must provide *delay-sensitive protocol stack*, take into account the delays caused by the *operating system (OS)* of the sensing equipment, and account for contention delays on the wireless channel.

In this section, we present an *integrated dynamic soft real-time (iDSRT)* solution [30] with *enforcement and adaptation mechanisms* that provides *statistical end-to-end delay* guarantees, but with much tighter jitter variance (spikes in delay performance) than in Section 4.2. Note that the reasons why the iDSRT solution is feasible is because (a) of the *controlled conditions* and environment of critical infrastructures (e.g., SCADA), where wireless (sensing) nodes do not join and leave in a random fashion; (b) of the *communication patterns* (who communicates with whom) which are well specified and understood; and (c) of the *limited scale* which assumes scale of sensor nodes in order of 10s and 100s of devices over WLAN enviroment with low throughput requirements in comparison to multimedia applications. However, even under these controlled conditions, introducing general purpose networking and computing technologies such as IEEE 802.11 WLAN and Linux/Windows OS is scary for the critical infrastructure providers. Hence, *intergrated QoS solutions* over these low-cost, but best effort, wireless technologies are needed.

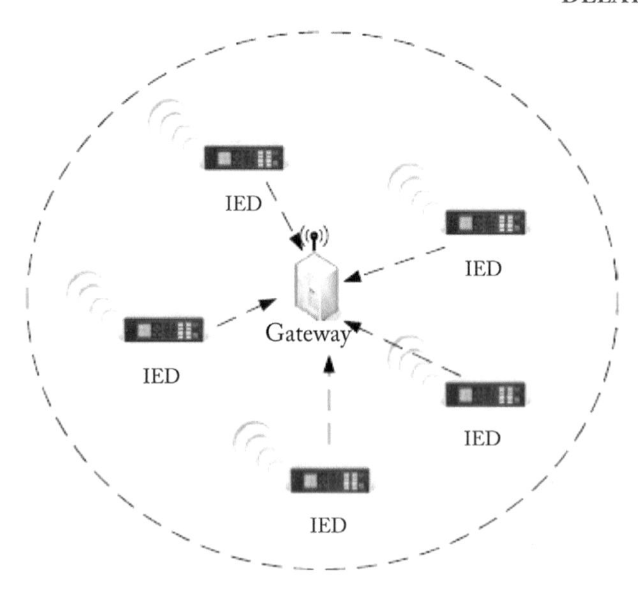

FIGURE 4.13: IEEE 802.11Wireless LAN deployment among wireless sensor devices such as the intelligent electronic devices used in power grid [30].

We will start with the presentation of the overall framework, followed by details of the dynamic scheduling algorithms, and finishing with discussion of practical considerations of this integrated QoS solution. Throughout Section 4.3, we will rely on the definitions and notation in Table 4.2.

TABLE 4.2: Notations and definitions for Section 4.3.

NOTATION	DEFINITION
$G = (V,E)$, $V = \{N_1, N_2,\ldots, N_n, S\}$, $E = \{(N_1, S), N_2, S), \ldots, (N_n, S)\}$	WLAN graph with n wireless sensory nodes and one gateway node S
n	Number of wireless sensor nodes
N_i	Sensory node i
m_i	Number of real-time tasks/applications in node N_i
$A_{ij}(C_{ij}, R_{ij}, P_{ij})$	j-th real-time task/application A_{ij} running on node N_i

(*continued*)

TABLE 4.2: (*continued*)

NOTATION	DEFINITION
P_{ij}	Period of real-time task/application A_{ij}
D_{ij}	Deadline of real-time application A_{ij} and $D_{ij} \leq P_{ij}$
C_{ij}	Number of time units A_{ij} consumes CPU resource to process its data (computation time)
PS_{ij}	Packet size of application A_{ij}
B_{ij}	Estimated bandwidth of wireless MAC of node N_i for application A_{ij}
R_{ij}	Number of time units A_{ij} sends one packet of size PS_{ij} with $R_{ij} = PS_{ij}/B_{ij}$ (transmission time)
$A_{ij}^{\text{CPU}}(C_{ij}, D_{ij}^{\{\text{CPU}\}}, P_{ij})$	CPU subtask of A_{ij} with computation C_{ij}, deadline D_{ij}^{CPU} and period P_{ij}.
$A_{ij}^{\text{Net}}(R_{ij}, D_{ij}^{\text{Net}}, P_{ij})$	Network subtask of A_{ij} with transmission time R_{ij}, deadline D_{ij}^{Net} and the period P_{ij}
$T_{ij} < D_{ij}; D_{ij}^{\text{CPU}} = T_{ij}; D_{ij}^{\text{Net}} = D_{ij} - T_{ij}$	Deadline assignment to the partitioned tasks of A_{ij} with $T_{ij} > 0$ and $T_{ij} < D_{ij}$, based on deadline assignment algorithm (Section 4.3.3)
$T = \{T_{ij} \mid T_{ij} > 0 \text{ and } T_{ij} < D_{ij}, i = 1,...,n; j = 1,...,m_i\}$	Deadline assignment for the whole task set in the system A_{ij}, $i=1,\ldots,n; j=1,\ldots,m_i$

4.3.1 iDSRT Architecture

To design the integrated framework, we assume the environment shown in Figure 4.13, where the wireless sensor devices and gateway are *computing devices* with general purpose operating systems such as Linux, supporting *multithreaded* and *networked applications*. Some of the application threads send sensory measurements and require real-time guarantees in terms of processing and packet delivery, some threads send management/control information with no real time requirements, and some of the application threads receive non-real-time control signals. These multithreaded and networked applications run in a WLAN environment, utilizing IEEE 802.11 among the devices and enabling the exchange of sensory measurement packets and control packets between sensory devices and the concentrator gateway. The traffic usually carries real-time-demanding measurement packets from the sensory devices to the gateway, non-real-time demanding control/management packets from sensors to the gateway and control/management packets from the gateway to the sensory devices.

Under the consideration of the general purpose wireless LAN network with wireless, static and general purpose computing nodes, we need to provide an *integrated* control of *three major resources* as part of the QoS solution. The resources are (1) *CPU* within individual computing nodes, (2) *network intranode bandwidth*, and (3) *network internode shared wireless channel* among wireless nodes. If these resources are not controlled in an integrated and coordinated fashion, they contribute to *end-to-end delay problems* of real-time-demanding packets from the sensory device to the gateway. When we discuss some of the practical issues, we will show the break-down of individual resource controls. On the other hand, since the resources in general purpose platforms run *best-effort resource management* solutions, e.g., IEEE 802.11 MAC functions, or Linux virtual memory management, which we do not control, we achieve only statistical delay guarantees via *dynamic and adaptive* control of resources. Hence, what we will be able to deliver in terms of QoS delay guarantees is statistical *soft-real-time guarantees.*

The integrated dynamic soft-real-time (iDSRT) framework consists of three major components, the *dynamic soft-real-time task scheduling* component (DSRT), residing in the kernel of each sensory node and scheduling application threads according to timing demands; the *network packet scheduling* component (iEDF), residing in the kernel of each sensory node and scheduling network packets according to timing demands; and the *distributed node coordination* component (iCoord), residing in the middleware layer/control plane of the sensory and gateway nodes and coordinating the nodes. The iDSRT framework architecture is shown in Figure 4.14.

The iDSRT system allows to run *real-time (RT)* and *best effort (BE)* networked applications (flows) together in one wireless device N_i, and it enables sharing of resources in a controlled manner. RT applications rely on *iCoord* which receives QoS specification (delay requirement) from RT applications, performs RT application profiling, and does the QoS negotiation on behalf of RT applications between devices N_i and the gateway. Its central role is managing resource allocation within each node and among nodes (sensors and gateway) to ensure end-to-end delay guarantees. Any potential conflicts among RT tasks and BE tasks on one node are resolved by the DSRT CPU scheduler. DSRT guarantees CPU resources for RT applications by using *adaptive EDF* (Earliest Deadline First) scheduling algorithm [32]. The guarantees are *soft* because DSRT does not manage other resources of the hardware and therefore does not prevent the preemptions due to non-CPU hardware interrupts. However, the soft timing guarantees are within requested application timing bounds. The *iEDF* component is the intranode and internode network packet scheduler that runs

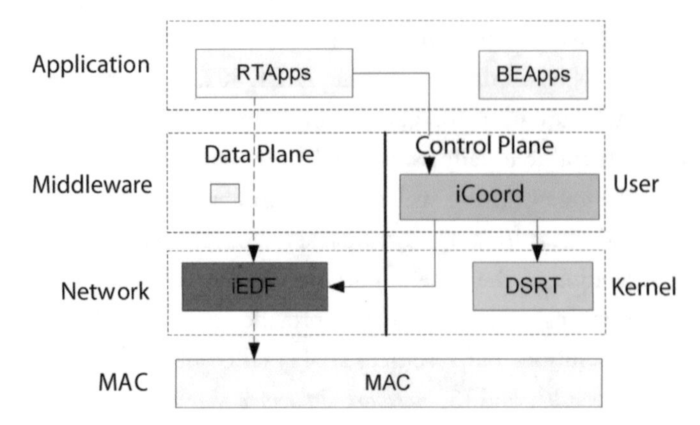

FIGURE 4.14: Integrated dynamic soft-real-time architecture for QoS delay guarantees [30].

Implicit Earliest Deadline First (iEDF) scheduling algorithm [33]. It takes the implicit contention approach to schedule transmission slots according to the EDF policy. It manages the packet queue of each node and makes sure all nodes agree on the same packet over the shared medium within a specific time slot.

4.3.2 Scheduling Components

As argued above, to achieve QoS delay guarantees within timing bounds in WLANs, it is important to consider nodes coordination, the CPU scheduler and the packet scheduler, that will cooperate with each other when executing individual resource allocations.

Coordinator (iCoord). This component is a distributed middleware component which coordinates all system scheduling components to ensure RT applications to meet their delay guarantees within timing bounds. It operates in the control plane of the node's protocol stack to provide the *node registration* service, *task profiling* and *coordination services*. Figure 4.15 shows the *middleware* iCoord *control architecture* as well as the *protocol* of iCoord. iCoord consists of two modules, the *Local iCoord* residing on each client (sensory node) and *Global iCoord*, residing on the gateway. Local iCoord is in charge of coordinating system components at each sensory node, and communicates with Global iCoord to assist in inter-node scheduling with other nodes Local iCoord(s). Global iCoord executes global services on the gateway.

Before any QoS guaranteed communication among sensory nodes and gateway can begin, each sensory node and its RT applications have to register with the iDSRT *registration service* through the Local iCoord Registrator. The registration request from a RT application includes (1) tuple of parameters such as the *process identifier, source address, source port, destination address and destination port*; (2) *period* of the RT application; and (3) RT application's *CPU cycles and network*

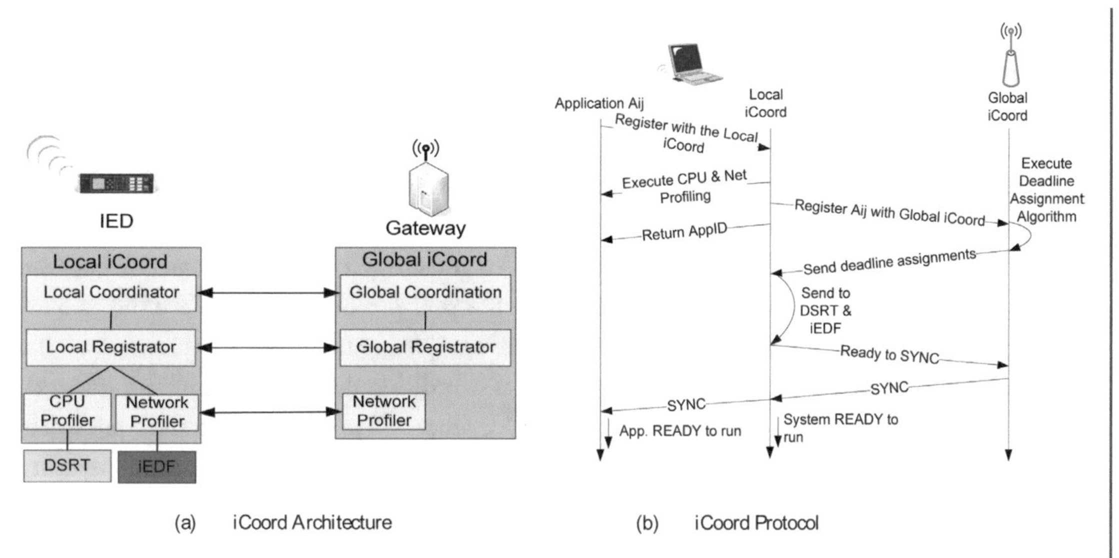

FIGURE 4.15: iCoord architecture (left) and iCoord protocol (right).

resource demands (pre-defined default values since the RT application does not have the demands at this point when registering). The Local iCoord Registrator sends the registration information for this application/node to the Global iCoord Registrator. After the Global iCoord Registrator acknowledges the successful registration, the Local iCoord Registrator returns a *unique ID* to the RT application, and invokes the CPU/network profiling services to approximate real CPU/network demands of the RT application. After profiling the RT application, the task profiles are sent to the Global iCoord Registrator for node admission control, inter-node scheduling and coordination to be performed at the gateway.

The *profiling services* of CPU and network resources are invoked at each sensory node after the registration phase, where the CPU usage (demand) is measured by having the task scheduler run several instances of the RT application task (thread/process). Similarly, network profiling is done by measuring the RT application packet round-trip time between the RT application sensory node and the gateway.

The *coordination* service includes the Local Coordinator at the sensory node and the Global Coordinator at the gateway. The Global Coordinator at the gateway gathers profiles of all RT applications from the Global Registrator and performs the *Deadline Assignment Algorithm*, discussed in Section 4.3.3. The assigned deadlines are then distributed to Local Coordinators. The information includes deadline assignments for the inter-node/intra-node (iEDF packet scheduler/DSRT task scheduler) scheduling of all tasks in the overall system. Upon receiving the deadline assignment of all tasks, the Local Coordinator confirms with DSRT and iEDF schedulers the acceptance of the deadlines. If acceptance is provided, Global Coordinator gets OK from all Local Coordinators and sends SYNC message for the delay-bounded transmission to start. Note that the BE transmission can flow anytime as long as it does not violate the RT applications (flows) demands.

Dynamic soft-real-time task scheduler (DSRT). The DSRT component is responsible for task scheduling according to their deadlines. To discuss the individual functions of the DSRT component, we need to present the RT application task model.

The *RT application task* model (see Figure 4.16.) takes into account the coordination between the EDF-based task scheduler and EDF-based network packet scheduler to achieve the end-to-end delay guarantees. It means if we denote a RT task as A_{ij} being the j-th RT application on the i-th sensory node (node N_i), with period P_{ij}, then we can divide the RT task into two subtasks, "compute" subtask A_{ij}^{CPU}, and "network" subtask A_{ij}^{Net}, processed in order due to their producer-consumer relation. It means within the period P_{ij}, the subtask A_{ij}^{CPU} needs C_{ij} time units for sensory sampling and processing data. After the data gets processed, the subtask A_{ij}^{Net} needs R_{ij} time units to send the measurement(s) packet to the gateway over the wireless network. In our task model we assume that the task deadline D_{ij} is the same as period P_{ij} (Note that the deadline D_{ij} is relative to the sensory node, i.e., by what deadline task A_{ij} needs to finish at the node N_i.). The deadline D_{ij} is also divided and we will have D_{ij}^{CPU} for the local deadline of A_{ij}^{CPU}, and D_{ij}^{Net} for the local deadline of A_{ij}^{Net}, where $D_{ij}^{\mathrm{CPU}} + D_{ij}^{\mathrm{Net}} = D_{ij}$. Both times C_{ij} and R_{ij} are calculated by the number of consumed cycles over the CPU frequency. Note that R_{ij} is the network transmission time of a packet transmitted by the task A_{ij}^{Net}'s network protocol stack. It transmits a packet of size PS_{ij} bytes over the wireless MAC with measured bandwidth B_{ij} at the i-th sensory node.

DSRT specifically manages real-time CPU tasks $A_{ij}^{\mathrm{CPU}}, j = 1,\ldots m_i$. To achieve this objective, DSRT is composed of four basic functions, the *Admission Control*, the *EDF-based SRT (Soft-Real-Time) scheduler*, the *Cycle Demand Adaptor*, and the *Overrun Protection*, as shown in Figure 4.17. The Admission Control checks if requested real-time tasks can be admitted, i.e., if there is enough CPU bandwidth to process real-time tasks by their deadlines. The SRT Scheduler schedules all accepted

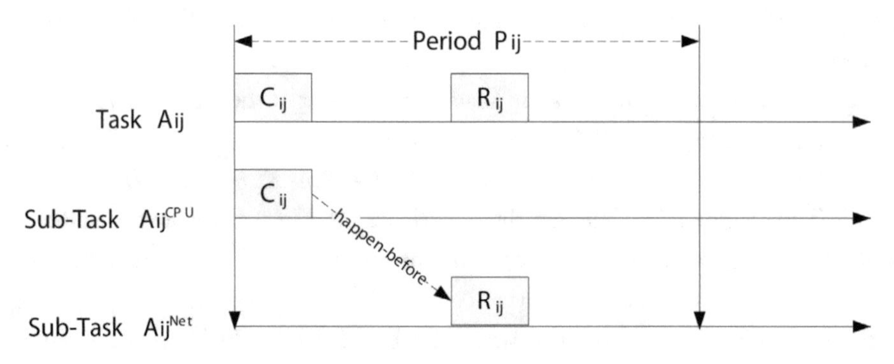

FIGURE 4.16: Real-time application model

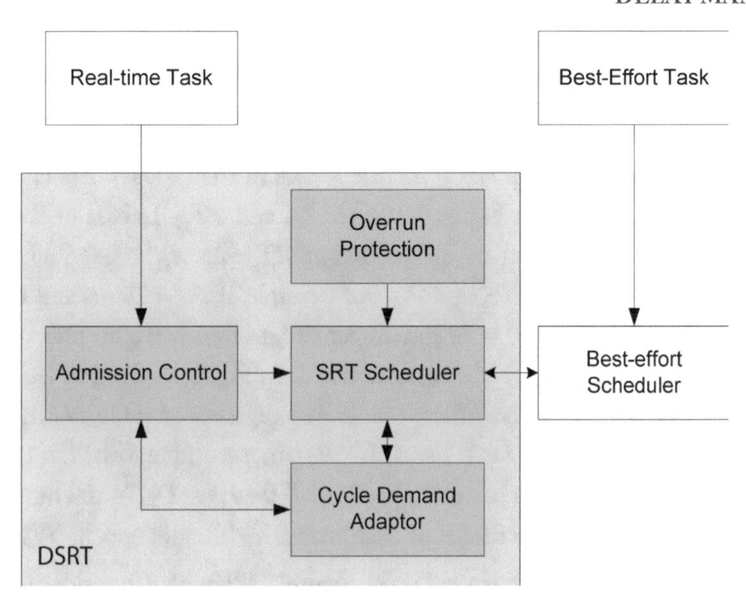

FIGURE 4.17: DSRT scheduling architecture.

real-time tasks according to the deadline-based policy, EDF (Earliest Deadline First). The Cycle Demand Adaptor tracks each real-time task and if it detects change (change in alloted/admitted number of cycles), it assists in adaptation of cycle allocation. The Overrun Protection module handles the situation when a real-time task cannot run in the alloted/admitted time. More details on each component are provided below.

Admission control. As discussed above, on each wireless node N_i, before using the real-time capabilities of the system, a new real-time (RT) task A_{ij}^{CPU} must register itself with *iCoord* as a RT task to be scheduled by DSRT. Specifically, it must specify its period P_{ij}, its worst case execution time C_{ij}, and its relative deadline D_{ij}^{CPU}. The CPU task admission control then uses the EDF schedulability test as follows:

$$\forall L \in \text{DLset}, L \geq \sum_{j=1}^{m_i} \left(\left\lfloor \frac{L - D_{ij}^{\text{CPU}}}{P_{ij}} \right\rfloor + 1 \right) C_{ij}, \tag{4.12}$$

where $\text{DLSet} = \{d_{kl} \mid d_{kl} = lP_{ik} + D_{ik}^{\text{CPU}}, 1 \leq k \leq m_i, l \geq 0\}$ is the set including all tasks' deadlines less than the hyper-period of all periods (i.e., least common multiplier of $P_{i1}, \ldots, P_{im_i}$). Equation 4.12 represents a *modified EDF scheduling condition*. Recall that the traditional EDF scheduling test would be $\sum_{j=1}^{m_i} \frac{C_{ij}}{P_{ij}} \leq 1$ [32]. However, since we have actually two subtasks (compute and network subtasks) in a RT task A_{ij}, the test needs to be refined towards the "compute" subtask A_{ij}^{CPU}. We will explain Equation 4.12. on examples, shown in Figure 4.18. Figure 4.18a shows one RT task A_{i1}^{CPU}

and its period and deadline parameters P_{i1}, D_{i1}^{CPU}. Figure 4.18b shows two RT tasks $A_{i1}^{CPU}, A_{i2}^{CPU}$ with parameters (P_{i1}, D_{i1}^{CPU}) and (P_{i2}, D_{i2}^{CPU}):

L represents the "compute" task deadline in each epoch for a task set m_i on a node N_i, where an epoch is equal to the *hyper-period HP of all task periods* in the task set. Epochs in Figure 4.18a are P_{i1}, $2{*}P_{i1}$, $3{*}P_{i1}$, $4{*}P_{i1}$. Epochs in Figure 4.18b are P_{i2}, and $2{*}P_{i2}$. In case of one RT task A_{i1}^{CPU}, the period P_{i1} is equal to the hyperperiod. In case of two RT tasks $A_{i1}^{CPU}, A_{i2}^{CPU}$, hyperperiod *HP* is the *least common multiple* (LCM) of P_{i1} and P_{i2}. For example, if $P_{i1} = 25$ ms and $P_{i2} = 50$ ms, then the hyperperiod HP = LCM(25, 50) = 50 ms. In general, to find *HP* is to find the least common multiple among all periods ($P_{i1}, P_{i2}, \ldots, P_{im_i}$) at node N_i. In Figure 4.18b, HP is P_{i2}.

L is less than the hyper-period of the considered m_i tasks at node N_i. In Figure 4.18a, L in the first epoch is $L = 0 * P_{i1} + D_{i1}^{CPU}$ and $L < P_{i1}$, in second epoch $L = 1 * P_{i1} + D_{i1}^{CPU}$ and $L < 2 * P_{i1}$, etc. In Figure 4.18b, L in the first epoch is $L = 0 * P_{i2} + D_{i2}^{CPU}$ and in the second epoch, $L = 1 * P_{i2} + D_{i2}^{CPU}$. *DLset* represents the set of deadlines L in different epochs '*l*'. In Figure 4.18(a), DLset $= \{D_{i1}^{CPU}, P_{i1} + D_{i1}^{CPU}, \ldots\}$. Similar set is for Figure 4.18b with L considering $m_i = 2$.

If we apply the schedulability test (Equation 4.12) for Figure 4.18a with $m_i = 1$ in the first epoch, we get

$$L \geq \left(\left\lfloor \frac{L - D_{i1}^{CPU}}{P_{i1}} \right\rfloor + 1\right) * C_{i1} = \left(\left\lfloor \frac{0 * P_{i1} + D_{i1}^{CPU} - D_{i1}^{CPU}}{P_{i1}} \right\rfloor + 1\right) * C_{i1} = (0+1) * C_{i1} = C_{i1}.$$

Hence, in the first epoch we have to satisfy schedulability test $C_{i1} \leq D_{i1}^{CPU}$, which is $\frac{C_{i1}}{D_{i1}^{CPU}} \leq 1$ and this clearly leads toward the form we know from the traditional EDF schedulability test for one real-time task in the system, only with tighter deadline D_{i1}^{CPU}. Similar calculation of schedulability test (Equation 4.18.) applies when we have two tasks and more. The term $\left(\left\lfloor \frac{L - D_{ij}^{CPU}}{P_{ij}} \right\rfloor + 1\right)$ in Equation 4.18 represents the *computational weight* for each execution time C_{ij}, depending on the *remaining time* between the overall deadline L in that epoch and the j-th task deadline D_{ij}^{CPU} relative to the j-th task period P_{ij}. For example, in Figure 4.18b if we consider the first epoch with $L = D_{i2}^{CPU}$, then Equation 4.18 will be

$$L = D_{i2}^{CPU} \geq \left(\left\lfloor \frac{L - D_{i1}^{CPU}}{P_{i1}} \right\rfloor + 1\right) * C_{i1} + \left(\left\lfloor \frac{L - D_{i2}^{CPU}}{P_{i2}} \right\rfloor + 1\right) * C_{i2} = \left(\left\lfloor \frac{D_{i2}^{CPU} - D_{i1}^{CPU}}{P_{i1}} \right\rfloor + 1\right) * C_{i1} + C_{i2}.$$

It means both subtasks $A_{i1}^{CPU}, A_{i2}^{CPU}$ must finish their execution before the deadline of the second task's deadline D_{i2}^{CPU}. Note that we need the weight to be at least 1 to account for the execution of each task as shown in Figure 4.18 examples.

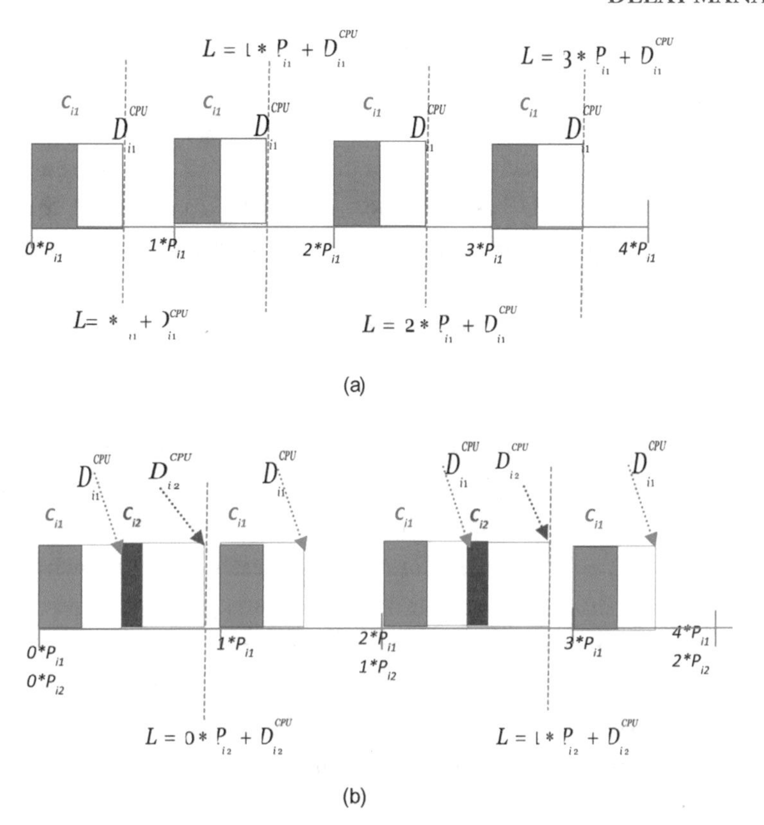

FIGURE 4.18: Examples of task schedules, deadlines, periods, execution times for (a) 1 RT Task A_{i1}^{CPU}, (b) 2 RT Tasks $A_{i1}^{\mathrm{CPU}}, A_{i2}^{\mathrm{CPU}}$ at node N_i.

SRT scheduler and overrun protection. If the condition Equation 4.18 is met, the task A_{ij}^{CPU} is added to the running queue of the *EDF Scheduler* and is scheduled to run in the next period. If the task cannot run in the alloted time C_{ij}, due to the demand cycle variation (note that the execution times C_{ij} are average times (cycles), and demand cycles for tasks can vary), the *Overrun Timer* will preempt the task to best-effort mode. In this case, the task A_{ij}^{CPU} will only be allowed to run after all other RT tasks have used their alloted CPU time. The Overrun Timer removes the task from the running queue and adds it to the *overrun queue*. Tasks in BE mode compete against each other and use the standard OS non-RT scheduler.

Cycle demand adaptor. Note that if the deadline D_{ij}^{CPU} is not met, the *Cycle Demand Adaptor* will assist in *adaptation* as follows. The Cycle Demand Adaptor tracks each RT application task A_{ij}^{CPU}. If it detects that the change in the cycle demand of A_{ij}^{CPU} is persistent, and D_{ij}^{CPU} cannot be met several times, it will aim to increase the alloted cycle demand for this particular task A_{ij}^{CPU}. It means the Cycle Demand Adaptor will query the DSRT Admission Controller if the increased alloted cycle demand of A_{ij}^{CPU} can be accepted. If the response is positive, the RT task A_{ij}^{CPU} gets an increased alloted time (C_{ij}).

Implicit earliest deadline first packet scheduler. iEDF is a distributed network scheduler that takes the implicit congestion approach to perform the EDF packet scheduling algorithm [33, 34]. Each sensory node N_i uses the iEDF as its network scheduler. Conceptually, this network packet scheduler is an *outgoing-packet scheduler*, working on top of the 802.11 MAC layer. It manages how packets are prioritized to ensure they will meet the deadlines.

iEDF is an implicit contention-aware scheduling algorithm that uses EDF for the packet scheduling. At any time slot, all sensory nodes agree on a RT task A_{ij}^{Net} to access the shared wireless medium according to the EDF policy. Specifically, for a node N_i, RT tasks $A_{ij}^{\text{Net}}, j = 1,\ldots, m_i$ are called *local RT network applications* and other RT tasks running on other nodes are called remote RT network applications. iEDF at each node N_i maintains the *deadline assignment* and the *task information of remote RT network tasks* in addition to its local RT network tasks, disseminated by iCoord. Once iEDF has all network task deadline information, it creates a "shadow network task" for each remote network task. The shadow network task has the same period, deadline and transmission time as the network task being shadowed. When the shadow network task $A_{kj}^{\text{Net}}, j = 1,\ldots,m_k$, "executes" at the sensory node N_i, $i \neq k$, it does nothing at N_i but sets up a time to wake up after the transmission time R_{kj} of task A_{kj}^{Net}. On waking up, the shadow network task again notifies iEDF that the remote task is supposed to finish. On this event, iEDF schedules another RT network task, either local or remote (shadow) for the next transmission. In this way, iEDF is doing a distributed EDF scheduling algorithm and packet collision rarely happens if each sensory node complies with the global deadline assignment for all RT network tasks in the system. Figure 4.19 shows the implicit contention scheduling.

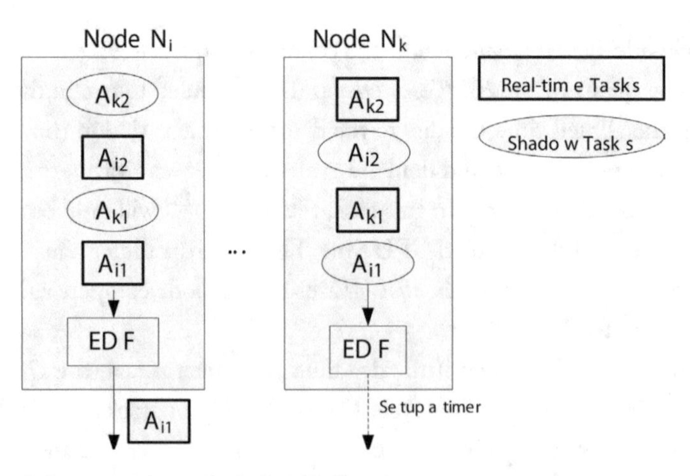

FIGURE 4.19: Implicit contention scheduling [30].

4.3.3 Deadline Assignment Algorithm

The deadline assignment very much depends on the employment of the EDF algorithm for the CPU scheduler (DSRT) and for the network packet scheduler (iEDF). Especially, the iEDF takes care of the packet scheduling (intranode scheduling) and the node scheduling (inter-node scheduling) via shadow network task consideration, assisted by iCoord and DSRT to achieve end-to-end QoS delay guarantees of RT applications A_{ij}.

We will illustrate the deadline assignment problem on the example in Figure 4.20. Let us consider the original RT task A_{ij} (C_{ij}, R_{ij}, P_{ij}) on a sensory node N_i, and its split into two subtasks, the CPU task $A_{ij}^{\text{CPU}}(C_{ij}, D_{ij}^{\text{CPU}}, P_{ij})$, and the network task $A_{ij}^{\text{Net}}(R_{ij}, D_{ij}^{\text{Net}}, P_{ij})$, where D_{ij}^{CPU} and D_{ij}^{Net} are relative deadlines for the CPU and network tasks with $D_{ij}^{\text{CPU}} + D_{ij}^{\text{Net}} = D_{ij} (\leq P_{ij})$. Furthermore, if the phase of the network task A_{ij}^{Net} is equal to D_{ij}^{CPU}, then the end-to-end delay of A_{ij} can be met as long as the CPU and network tasks meet their deadlines with the node N_i.

A deadline assignment $T = \{T_{ij} \mid T_{ij} > 0 \text{ and } T_{ij} < D_{ij}, i = 1,...,n; j = 1,...,m_i\}$ is a set of deadlines T_{ij} assigned to each corresponding RT task A_{ij}, i.e, $D_{ij}^{\text{CPU}} = T_{ij}$ and $D_{ij}^{\text{Net}} = D_{ij} - T_{ij}$. T is valid if it yields a feasible scheduling for the CPU task set at each node and the network task set in the system. Furthermore, in addition to the validity constraint, T can be optimized according to an *objective function*. Different objective functions lead to different ways to assign deadlines and different solutions [35, 36, 37, 38].

It means, a solution of the deadline assignment problem T specifies a set of $\{T_{ij}\}$, $i = 1,..., n$, and $j = 1,..., m_i$, where (1) assigning $D_{ij}^{\text{CPU}} = T_{ij}$ and $D_{ij}^{\text{Net}} = D_{ij} - T_{ij}$ will yield a feasible scheduling for all CPU and network tasks in the system, and (2) T is optimized according to an objective function. Intuitively, for any deadline assignment T, decreasing T_{ij} will put more stress on the CPU of node N_i, and less stress on the network, and vice versa. Thus it is desired to put stress fairly on both resources.

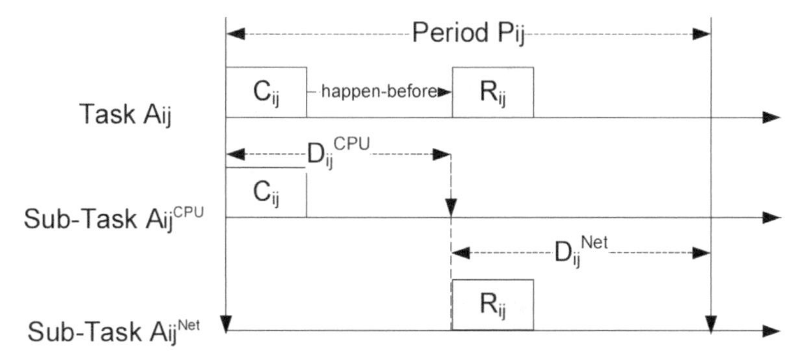

FIGURE 4.20: Illustration of deadline assignment problem on task A_{ij} [30].

Formally, we define the *stress factor* on CPU resource of a deadline assignment T_{ij} as $F_{ij}^{\text{CPU}}(T_{ij}) = \frac{C_{ij}}{T_{ij}}$, the stress factor on the network resource $F_{ij}^{\text{Net}}(T_{ij}) = \frac{R_{ij}}{D_{ij} - T_{ij}}$, and the total stress of a deadline assessment T_{ij} as $F_{ij}(T_{ij}) = F_{ij}^{\text{CPU}}(T_{ij}) + F_{ij}^{\text{Net}}(T_{ij})$. It is important to notice that due to the *dependency* among subtasks (producer-consumer dependency), the resource partititioning always puts *more stress* on both resources.

Thus, it is preferable to *minimize* the total stress of all tasks in the system over all possible deadline assignments T:

$$\min F(T) = \sum_{i=1}^{n} \sum_{j=1}^{m_i} F_{ij}(T_{ij}), \forall T$$

$F(T)$ is used as the objective function to the deadline assignment problem.

The *constraints* to this optimization problem come from the CPU task and network packet scheduling. The CPU task and network schedulers need EDF admission control, where tasks have deadlines less than their periods. This can be done by using the *processor demand criteria* [39, 40]. The overall problem can be formulated as an optimization problem:

min $F(T)$
Variables $T = \{T_{ij}, i = 1,\ldots,n; j = 1,\ldots,m_i)$
Constraints $C_{ij} < T_{ij} < D_{ij}, i = 1,\ldots,n; j = 1,\ldots,m_i$
Schedulability tests for CPU and network tasks

The above optimization problem is a *non-linear optimization problem* because the objective function is *nonlinear*. However, the objective function is a convex function because it is the sum of convex functions $F_{ij}(T_{ij})$ [41]. Furthermore, except the EDF-based schedulability tests for CPU and network tasks, all other constraints are linear. Hence, the overall optimization problem is a *convex optimization problem* and using standard convex optimization solving techniques such as Lagrange multiplers method [41], the problem can be solved theoretically and numerically. Once solved, the result is the deadline assignment $T = \{T_{ij}, i = 1,\ldots,n; j = 1,\ldots,m_i\}$ assigned to each corresponding task A_{ij} (i.e., $D_{ij}^{\text{CPU}} = T_{ij}, D_{ij}^{\text{Net}} = D_{ij} - T_{ij}$).

4.3.4 Practical Issues

Even though the principles of the coordinated scheduling are straight forward, several issues need to be addressed in the practical system.

First, the network task scheduling depends on the correct estimation of the transmission time of the shadow network task. For any particular transmission, the remote task A_{kj}^{Net} may finish earlier than expected due to worst case profiling and estimation of R_{kj}. It may also finish later than expected due to the noisy and unreliable channel. In the former case, iEDF ignores the early transmission and allows BE (Best Effort) task transmission. In the later case, iEDF has to avoid starting another transmsision to minimize packet collisions. To resolve this issue, iEDF goes to the listening mode and overhears the wireless network to know when the remote network task finishes.

Second, the iEDF network packet scheduler consumes non-negligible CPU resource for scheduling. To resolve this issue, we charge the network task's scheduler CPU consumption to the computation time of the corresponding RT application task.

Third, as we mentioned briefly above, one needs the coordination of all three resources, the CPU, intra-node network and inter-node network resources, to achieve *end-to-end delays* within desired timing bounds as shown in Figure 4.21. The results show the end-to-end delay performance for iDSRT as well as for deploying only CPU scheduler DSRT, only iEDF network scheduler, and only BE system. Note that in the iEDF case, we have only the intra-node EDF-based packet scheduling, i.e., no internode scheduling, no distributed coordination and no DSRT scheduling. The results have been obtained in a testbed with 7 laptops IBM T60 Dual Core 1.66 GHz, IEEE 802.11 a/b/g Atheros-based wireless card (results use IEEE 802.11a mode), Linux kernel version 2.6.16

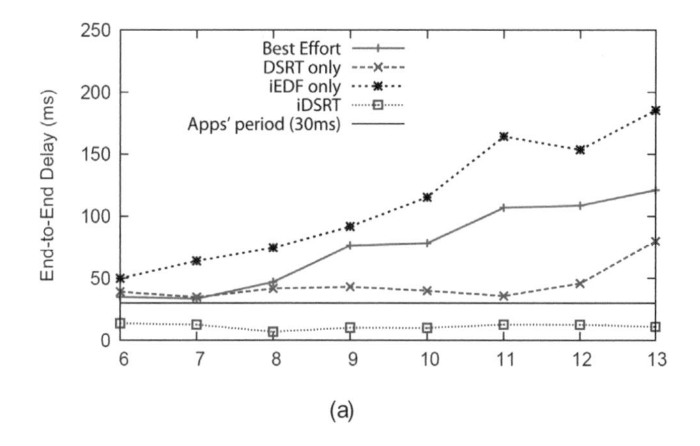

(a)

#Apps	Best Effort	DSRT-only	iEDF-only	iDSRT
6	90.06	90.54	51.58	34.17
7	93.75	88.40	69.90	35.38
8	96.88	92.34	82.82	35.59
9	482.05	101.30	102.43	33.78
10	508.31	109.15	138.21	33.59
11	505.76	97.77	154.13	34.17
12	516.69	128.25	164.42	34.63
13	558.37	136.11	169.35	35.25

(b)

FIGURE 4.21: Average end-to-end delays (left) and maximum end-to-end delays in ms (right) [30].

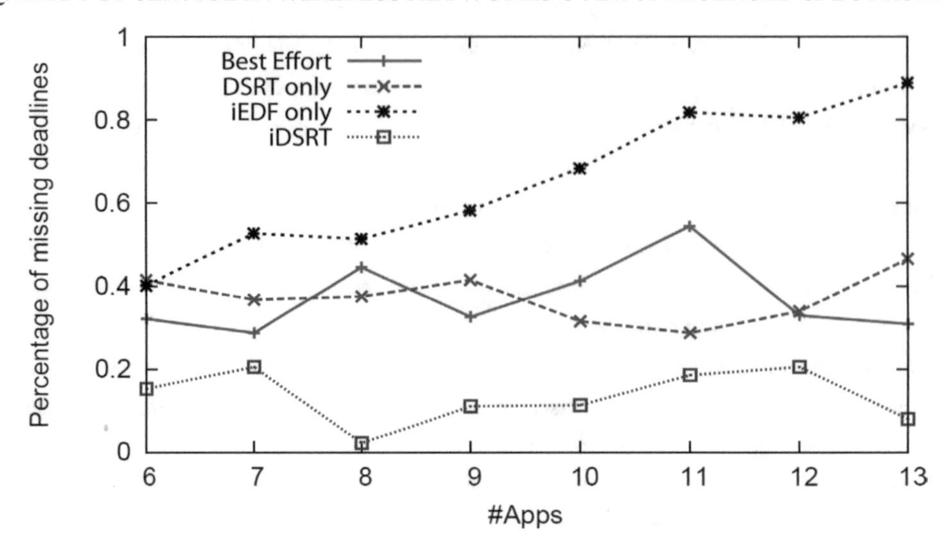

FIGURE 4.22: Missing deadlines [30].

with high resolution timer patch. The laptops were configured according the topology, shown in Figure 4.13., with one laptop serving as the gateway and the others as sensory nodes. The network operated on the channel that had least interference to minimize external effects and emulated the critical infrastructure setup.

Fourth, another important metric to consider within these types of critical infrastructure WLAN systems is the *missing deadline* of RT application tasks. As results in Figure 4.22 show (in the same testbed as above), iDSRT misses low percentage of deadlines of RT application tasks, where if employed only iEDF packet scheduling, the performance is truly bad.

Note that missed deadlines mean losses in RT applications since a late packet is a lost packet. Hence the losses due to the missed deadlines can be high if no QoS solutions are deployed.

4.4 SUMMARY

In this chapter we have shown QoS solutions for achieving end-to-end delay guarantees that included *adaptive* and *integrated* approaches. As it was pointed out in Chapter 1, deterministic QoS are not possible to achieve in wireless networks, hence to come close to desired end-to-end QoS guarantees (a) several layers must cooperate together, and (b) adaptations must be in place to adjust the resource allocation.

In Section 4.2, we have shown the importance of upper layers such as application and middleware layers as well as cross-layering with lower layers in the overall end-to-end QoS provision. Especially, we have shown the impact of upper layer adaptations on the overall QoS solutions. It is important to stress that applications have a rich set of contextual information about the semantics of the transmitted data and this semantic information allows for much better data differentiation than the underlying layers in case of network resource contention. Deploying upper layer adaptations such as PI controllers can nicely keep the end-to-end delays steady within probabilistic bounds as shown in Section 4.2.

In Section 4.3, we have shown the importance of integration, coordination and adaptation among OS *CPU resource* control, *network intranode* control, and *network internode control* when aiming to achieve QoS solutions such as *end-to-end delay guarantees*. Many real-time solutions exist within OS and network domains, and often QoS results for individual layers and subsect of cross-layers are shown within OS or within IEEE 802.11 wireless networks. However, we want to stress that *integration* of delay-sensitive solutions in a coordinated manner is needed to yield the desired QoS solutions as the Section 4.3.4 shows.

The next important issue for complete QoS solutions is *routing* that represents an integral component of the end-to-end QoS puzzle in wireless multi-hop networks. Chapter 5 will provide some insights of impact of routing on QoS provisioning.

• • • •

CHAPTER 5
Routing

5.1 INTRODUCTION

Diverse wireless networks are becoming an integral part of the ubiquitous computing and communication environment, providing new infrastructure for multiple applications such as video phone, multimedia-on-demand and others. In order to access multimedia information, certain level of Quality of Sertvice (QoS) needs to be considered such as high success ratio to access multimedia data, bounded end-to-end delay, low energy usage, high bandwidth rate, and others. We have discussed some of the QoS and resource enforcement and adaptation techniques in Chapters 3 and 4, especially at the end-nodes to enable efficient multimedia data delivery. One function that remains to be discussed is the *routing function* that connects sources and destinations of information when it moves information/packets through the mobile or static multi-hop/ad hoc wireless network.

Before we discuss the various routing aspects in more details, we present the routing function with its ties to higher layer functions via *cross-layer design*. Routing functions, similar to scheduling and rate allocation functions in Chapters 3 and 4, benefit from the interaction between the middleware and routing layers. We will briefly show the cross-layer interactions and benefits on an example.

Let us consider the cross-layer system architecture between application, middleware and routing layers in Figure 5.1 [54]. The cross-layer architecture includes (1) group-based application that produces and shares multimedia data with other users within a group in the network; (2) *middleware* that runs the data accessibility service to assist applications in locating, accessing and replicating data; and (3) *routing* that computes feasible routes and forwards packets to other mobile nodes within a group in the network. The sharing between routing and middleware layers is done in the form of *system profiles*. For example, system profiles may include contextual information useful such as nodes' *location* and *motion* pattern, known at the routing layer, and *data priority* information, known at the middleware layer. Middleware utilizes nodes' location and movement pattern to predict future connectivity of a group for data replication purpose. Routing utilizes data priority to differentiate network packets for routing purposes.

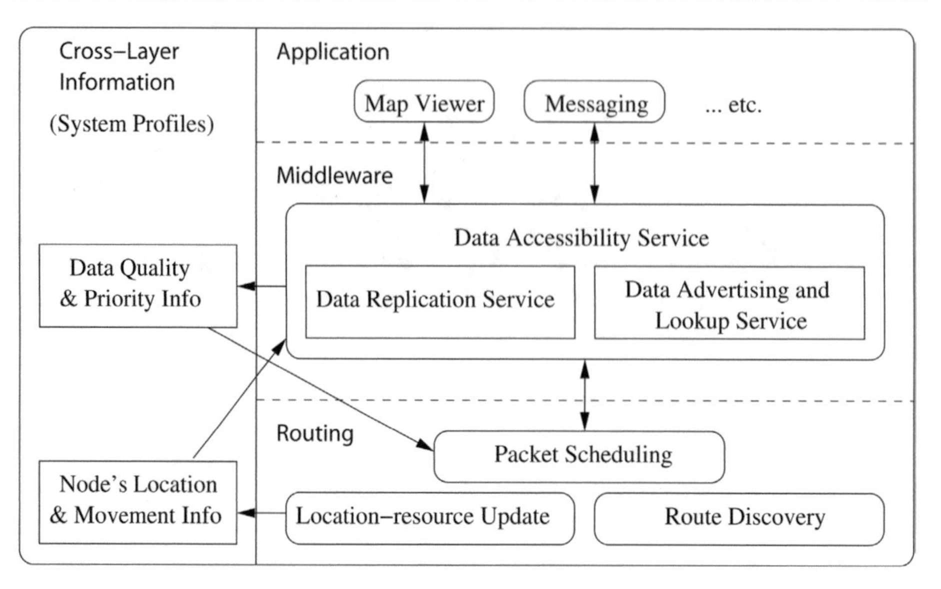

FIGURE 5.1: Cross-layer system framework between middleware and routing layers [54].

The benefit for routing function from the cross-layer design is manifold. The routing function yields (a) *better routing results* in terms of route discovery, selection and maintenance, and (b) *less computational and communication overhead* due to access to higher level information such as nodes' location and mobility patterns.

There is a huge body of wireless routing work, hence what we present here is only an exemplary cut through possible routing designs and solutions which might provide guiding directions to the reader, i.e., where to start when facing certain wireless computing and communication environments. Overall, the wireless routing space can be characterized and classified along many dimensions:

(a) routing over *mobile* vs *static* multi-hop ad hoc networks [e.g., 42, 47, 52, 53, 54];
(b) routing with *QoS support* vs *best effort* [e.g., 47, 48, 57, 58];
(c) *point-to-point* routing vs *multicast* routing [e.g., 42, 47, 54, 59, 60];
(d) *network* layer routing vs *overlay* routing [e.g., 42, 54, 59, 60];
(e) *source* routing vs *distributed* routing [e.g., 42, 47, 54];
(f) *proactive* vs *hybrid* vs *reactive* mobile ad hoc routing [e.g., 48, 50, 51, 74, 102, 123];
(g) routing for networks with *persistant* wireless connectivity vs routing for delay-tolerant networks with *intermittent* wireless connectivity [e.g., 42, 47, 54, 61];
(h) routing considering *user-based* optimization with their social structure vs routing considering *system-based* optimization [e.g., 43, 44, 45, 46];

(i) routing considering *good* paths vs *bad* paths caused by malicious and misbehaving nodes [e.g., 47, 49, 50, 51, 54];

(j) routing with route optimization according to different *QoS metrics* such as low energy, low end-to-end delay, high throughput, high delivery rate, low loss rate [e.g., 42, 43, 44, 45, 46, 54];

(k) routing under different *mobility models* [e.g., 55, 56];

(l) routing with and without the aid of *location* information [e.g., 52, 53, 59, 60] and other contextual information.

In this chapter we will discuss three routing algorithms taking into account QoS requirements and optimizing their routes along multiple routing dimensions. We aim to present selective routing examples that cover some of the routing dimensions, but by all means, the routing topic is not covered comprehensively due to the very large scope. We will show routing issues ranging from algorithmic and protocol design, inclusion of routing into a cross-layer system framework, cooperation with middleware layer, to theoretical QoS-resource optimization considerations. Our goal is to show the different routing considerations that one needs to take when given requirements, as well as what questions to ask. All three of these routing algorithms show (1) *very different design choices* and approaches to find feasible routes and (2) the *diversity* of the overall routing problem.

5.2 PREDICTIVE LOCATION-BASED QoS ROUTING

We present the *predictive location-based QoS routing protocol* [54] that assists *application/middleware services* to provide *access to multimedia* information for a *group* of mobile users with statistical *QoS requirements* such as high success rate, bounded end-to-end delay. An example of such application/middleware services was discussed in the Section 5.1.1. The predictive location-based QoS routing will consist of two functions, the *location prediction*, and the *QoS routing*. In Section 5.2.1, we discuss the location prediction function with (1) the *location-resource update protocol* to distribute location/resource information and provide route information to be used in *route discovery*, and (2) the *location prediction scheme* to estimate new location at a future instant, based on information regarding previous locations and previous end-to-end delays, and provide route information to be used in *route selection*. In Section 5.2.2, we describe the *QoS routing protocol* to route multimedia data in an efficient and high quality fashion.

Overall, the presented *routing algorithm* [54] covers some of the characteristics, discussed in Section 5.1.2, such as it (a) works on *mobile ad hoc networks*, (b) optimizes *end-to-end delay* for multimedia data access, (c) utilizes *location* of nodes for routing, (d) is part of a *cross-layer design*, (e) is *proactive source routing*, and (f) is *point-to-point network layer* routing approach.

Throughout the Section 5.2 we will use notation and definitions shown in Table 5.1.

TABLE 5.1: Notations and definitions for Section 5.2.

NOTATION	DEFINITION
F	Type-1 update frequency of geographic and resource information; usually weakly periodic update frequency
$f_{\max}$	Maximal threshold of type-1 update frequency
$f_{\min}$	Minimal threshold of type-1 update frequency
V	Velocity (speed) of a node
v_c	Velocity (speed) at location (x_c, y_c)
Δ	Threshold distance between actual (expected) and predicted location
t_c	Periodic check time point for type-2 update
(x_e, y_e)	Expected location of a node
(x, y)	Geographical location of a node
(x_c, y_c)	Current location of a node
t_p	Future time point at which location is to be predicted
(x_p, y_p)	Predicted location of a node
a, b	Nodes; a is source and b is destination

5.2.1 Location Prediction

The network level approaches for mobile nodes such as the *location-resource update* protocol and *location prediction scheme* need to assist the overall QoS routing.

Location-resource update protocol. The *update protocol* is crucial for distribution of geographic location and resource information. We consider resources such as *battery*, *power*, *queueing space*, *processor utilization*, *transmission range*, etc. In this protocol, we assume that all clocks in the mobile ad hoc nodes are synchronized, hence all update information is properly ordered. We also assume that the geographic location is obtained via GPS or similar location determining mechanism. The *update protocol* will use *two types of updates*:

- *Type-1 update* which can be generated in a weakly periodic manner. It can be generated with a constant frequency, i.e., the time between successive type-1 updates remains constant, or the frequency of type-1 update can vary linearly between maximum ($f_{\max}$) and minimun ($f_{\min}$) threshold depending on the velocity v of the mobile node. Consequently, the distance traveled between successive type-1 updates remains either constant or varies within a certain bound. This function is shown in Figure 5.2.
- *Type-2 update* is generated when there is a considerable change in the node's velocity or direction of motion. From its recent history, the mobile node can calculate an expected location that it should be in at a particular instant. The node then periodically checks if it has deviated a distance greater than a distance δ from its expected location. If it had deviated more than a distance δ from its expected location, a type-2 update is generated.

The *algorithm* to generate the *type-2 update* is as follows:

Suppose that the periodic check for a particular node is scheduled at time t_c. Then the node finds whether it has deviated more than a distance δ from its expected location (x_e, y_e) at t_c. Further, let us suppose that its own most recent update was generated at time t, $t = t_c - \Delta t$, where Δt is some time interval. Let us assume that this update was generated at point *(x,y)*, reporting velocity v, and direction such that the node moves at an anti-clockwise angle θ to the horizontal axis. Also, let us assume the current location of the node at the time of checking t_c is (x_c, y_c). Let the velocity at the time of checking v_c remain unchanged since the last update, i.e., $v_c = v$. This situation is shown in Figure 5.2b. Then, the expected location (x_e, y_e) is as follows: $x_e = x + v \cdot (t_e - t) \cdot \cos\theta$ and $y_e = y + v \cdot (t_e - t) \cdot \sin\theta$. Now if the *condition* $[(x_e - x_c)^2 + (y_e - y_c)^2]^{1/2} > \delta$ is satisfied, then a *type-2 update is generated* at the time t_c of checking due to significant change in motion pattern. It

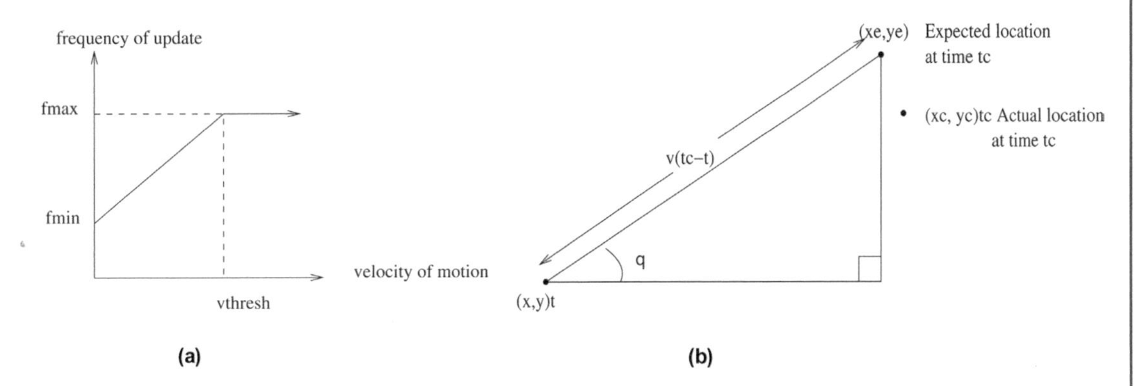

FIGURE 5.2: (a) Variation of update frequency of type-1 update with velocity of the node. (b) Type-2 update generation check [54].

is important to take care of the value δ to make sure that it is large enough to prevent the reporting of minor perturbations in direction. Alternatively, if there is a significant change in the velocity at t_c, then a type-2 update is generated due to significant change in the motion pattern.

Update packets contain *timestamps, current geometric coordinates, directions of motion (optional), velocity and other node resource information* that is used in QoS routing. Updates also contain a *motion stability parameter*, which is used in QoS routing. The *motion stability parameter* is a special QoS parameter, represented by a single bit, that indicates whether the update has been generated by type-1 or type-2 updates. Conceptually, the motion stability parameter indicates *whether the velocity and direction of motion are constant* and predictable or dynamically varying. If the velocity and direction are dynamically varying, they are hard to predict accurately and are not good candidates to be used as intermediate nodes for connections requiring low delay and low delay jitter. Using such nodes with unstable motion patterns as intermediate nodes results in *frequent re-routing* which increases delay and delay jitter. The update packet also has a field for the mobile node to indicate whether it is moving in a *piece-wise linear pattern* or *angular pattern*. Updates from a node in the network are propagated to other nodes by broadcast flooding similar to DSDV (Destination-Sequenced Distance Vector) routing [74] and DREAM (Distance Routing Effect Algorithm for Mobility) [75].

Location prediction scheme. When a packet arrives at a node a to be routed to a particular destination b, node a has to follow a *two-step process* to forward the packet along.

- The first step is to predict the *geographic location* of the destination b as well as the *candidate next hop nodes*, at the instant when this packet will reach the respective nodes. Hence, this step involves a *location* as well as *propagation delay prediction*. The *location prediction* is used to determine the *geographic location of some node* (either an intermediate node or the destination b) at a particular instant of time t_p in the future when the packet reaches it. The *propagation delay prediction* is used to estimate the value of t_p (to be used then in the location prediction). These predictions are performed based on previous updates of the respective nodes.
- The second step is to perform *QoS routing* based on the candidate *next hop node* information, determined in the first step, as well as the *resource availability* information for the candidate next hop nodes.

At this point we will describe the *location prediction* and the *propagation delay prediction* schemes.

Location prediction. We derive this scheme based on the assumption that a node moves in a *piecewise linear pattern*, i.e., between successive update points, the node has moved in a straight line. For a piecewise linear motion pattern and update packets that do not contain direction information, two previous updates are sufficient to predict a future location of the mobile node in the plane as follows:

Let (x_1, y_1) at t_1 and (x_2, y_2) at t_2 $(t_2 > t_1)$ be the latest two updates, respectively, from a destination node b to a particular correspondent node a. Let the second update also indicate v to be the velocity of b at (x_2, y_2). a wants to predict the location (x_p, y_p) of node b at some instant t_p in the future. The situation is shown in Figure 5.3. The value of t_p is set by a to current time + predicted delay for the packet to reach b from a. Then, using the similarity of triangles (as shown in Figure 5.3), we get $\frac{y_2 - y_1}{y_p - y_1} = \frac{x_2 - x_1}{x_p - x_1}$. From this similarity equation, we can calculate y_p if we know x_p:

$$y_p = y_1 + \frac{(x_p - x_1) \cdot (y_2 - y_1)}{x_2 - x_1}.$$

We can calculate x_p using Pythagoras' theorem $(x_p - x_2)^2 + (y_p - y_2)^2 = v^2(t_p - t_2)^2$ and another similarity of triangles for $y_p - y_2$ from Figure 5.3, getting

$$x_p = x_2 + \frac{v(t_p - t_2)(x_2 - x_1)}{[(x_2 - x_1)^2 + (y_2 - y_1)^2]^{1/2}}.$$

In the case of updates that include information about the direction of motion of the nodes, only one previous update is required to predict a future location. The calculation of the predicted location at a correspondent node a is then exactly the same as the periodic calculation of the expected location (x_e, y_e) at the node b itself.

Delay prediction. One possible scheme is to use the most recent update coming from the destination b, if we assume that the end-to-end delay for a data packet from a to b will be the same as the delay experienced by the latest update from b to a. In Figure 5.3, the delay prediction then will be

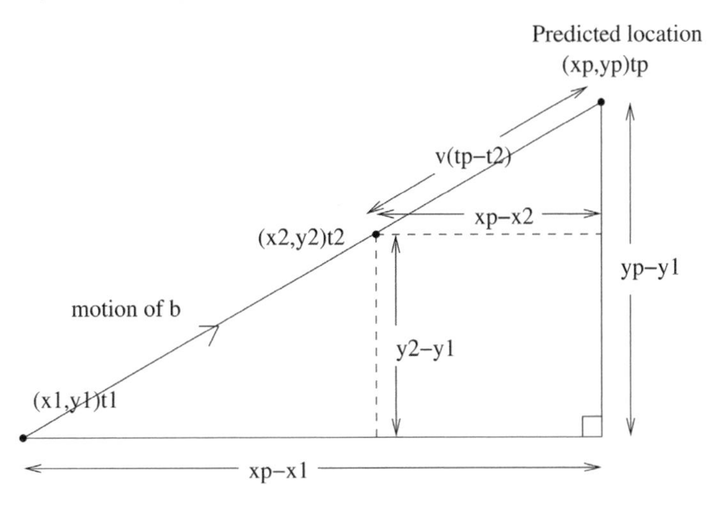

FIGURE 5.3: Future location prediction of node b at future time t_p using last two updates at time t_1 with update (x_1, y_1) and at time t_2 with update (x_2, y_2) [54].

$t_2 - t_c$, where t_2 is the timestamp of node *b* when it sends the most recent update (x_2, y_2) to node *a*, and t_c is the current time at node *a* (node *a* is not marked in Figure 5.3.). It means the end-to-end delay prediction to node *b* will be equal to the delay experienced by the *most recent update* that has arrived from *b*. Note that this is feasible since we assume that clocks of mobile nodes are synchronized.

5.2.2 QoS Routing

The location-and-delay prediction is extensively used in the QoS routing decision. The presented routing algorithm is a *proactive source-driven routing* algorithm, where each node *a* has information about the whole topology of the group network. The reason for the predictive location-based QoS routing protocol being a *table-driven, proactive source routing* is the *group-based application class* that we assume (see Figure 5.1). In this class of applications, to enable seamless, fast and high quality access and browsing of data, we need sources to predict the data locations, route fast and access data in efficient manner. Hence, we need a proactive sharing of routing information accessible to sources. One could consider also other routing algorithms such as reactive source-driven or distributed routing algorithms, but the access time to shared data would suffer due to higher overhead in route discovery and route selection. A proactive source routing works best if (a) nodes move slowly, and/or (b) nodes move in coordinated fashion, and/or (c) nodes move in predictive (e.g., regular) fashion which is the case in the group-based application.

The proactive source routing can thus compute the route from itself to any other node, using the routing information it has, and can include this source route in the packet to be routed within a group. The state information about the group network at a node *a* consists of two tables—the *update table* and the *route table*.

The *update table* contains information related to every node that *a* receives updates from. The update table stores the *node ID* of *b*, the *time the update* packet was *sent*, the *time of update* it was *received*, the *geometric coordinates* of *b*, the *speed* of *b*, the *resource parameters* of *b*, and optionally, the *direction of motion* of *b* as contained in the update packet. In the update table, the entry for some node *k* also contains *proximity list*, which is a list of all nodes lying within a distance of 1.5 × transmission range (called proximity distance) of *k*. The reason for choosing 1.5 × transmission range is because we assume that a node, which is not even within the proximity distance of *k* at the update reception time *t*, would not have moved within the transmission range at route computation time $t + \delta t$. Furthermore discussion follows in Section 5.2.3.

The proximity list is used at the time of *QoS route computation (selection)*. During the route computation, the source *a* is required to compute the neighbors of the intermediate node *k*, which are within *k*'s transmission range to be used as next hops from *k*. However, if we only maintain a *neighbor list* for *k* rather than *proximity list*, then nodes that were outside of *k*'s transmission range at the time of their respective last updates, but have since moved into it, will not be considered.

The update table stores the *last* 2 (for piecewise linear motion) or 3 (for angular motion) updates of *b* for the location-and-delay predictions.

The *routing table* at a node *a* contains all routes originating at *a*. When an update is received at a node *a*, it checks if any of the routes in its route table is broken or is about to be broken to see if *route re-computations* need to be initiated. Owing to the location prediction scheme based on the updates, we can predict if neighboring nodes on a route are about to move out of each other's transmission range. Thus, route re-computations could be initiated before the route actually breaks.

The *QoS routing algorithm* at the network layer has then the following steps to provide the *route discovery, selection, and maitenance* functions:

1. When a connection request with QoS requirements arrives from the middleware layer to the network level of node *a*, it first runs a *location-delay prediction* on each node in its proximity list, utilizing the update table, and thus obtains a list of its neighbors' nodes locations and delays at the current time.
2. It finds out which of these neighbors have the resources—*maximum delay, required stability* and *battery power*—to satisfy QoS requirements of the connection/route. We call these neighbors that satisfy the QoS requirements "*candidates.*"
3. It then performs a *depth-first search* for the destination starting at each of these candidate neighbors to find all *candidate routes*. At each step of the depth-first search, only the neighbors with resource levels satisfying the QoS requirements of the connection are considered. If the source has mistakenly assumed a node's physical proximity to another to imply connectivity between them, that node sends back a gratuitous update indicating that this assumption is incorrect.
4. From the resulting candidate routes, the *geographically shortest* one is selected as the route to forward the packet along. The packet is then forwarded along this route to the destination. The source route is included in the packet.
5. After connection establishment, packets are forwarded along the computed source route until the end of the connection or until the route is *recomputed* due to breakage or in anticipation of breakage.

5.2.3 Practical Issues

There are several issues to consider as one designs routing approaches for *groups* operating in MANET environments.

First, the above QoS routing scheme *does not reserve* resources at intermediate nodes in a connection and *does not provide deterministic QoS guarantees*. As we mentioned previously, in a network as dynamic as a *mobile ad hoc network*, it is not feasible to reserve network resources for an end-to-end connection, and consequently *not possible to provide deterministic QoS guarantees*. The *cross-layer* design can provide some QoS-aware admission control and also perform *predictive route adaptation* to adjust to changes in network resources that affect QoS at the application level. But, at best one can provide *soft, per-flow statistical QoS guarantees* of resource availability and connection stability.

Second, various design choices were made which have often *practical justifications*. For example, we choose *1.5 transmission range* as the *proximity distance* because we assume that a node, which is not even within distance 1.5 × transmission range of k at the update reception time t, would not have moved within the transmission range at route computation time $t + \delta t$. Thus, based on node b's location update, node a inserts the update into the proximity list of all nodes k in its update table that are reachable within 1.5 × transmission range of node b, at the time of reception of the node b's update.

Third, the above discussed update protocol involved *flooding* of *location* and *resource information* pertaining to a node to all other nodes in the group network. Ordinarily, such a full flooding of the network involves a *very large overhead*. However, with schemes that impose a hierarchy on the mobile ad hoc network, such as the MPR (Multi-Point Relays) scheme [76] or WHIRL (Wireless Hierarchical Routing Protocol) [80] for group mobility, the overhead associated with flooding can be considerably reduced. For example, in the MPR scheme, certain nodes are elected as multi-point relays (MPRs) for their neighborhoods. Nodes that are not MPRs receive and process the flooding message from their neighborhood MPRs, but do not re-broadcast it. Only the designated MPRs re-broadcast the flooded message.

Forth, the presented routing approach is heavily based on *location and delay predictions* and the goodness of these predictions is measured by the *accuracy of the prediction*, i.e., by the *prediction error metric* to gauge the accuracy of our prediction mechanism. Figure 5.4 shows the *accuracy of location and delay predictions*. The *prediction error* is defined as the percentage of predictions that were off the actual location by a particular distance. The interesting aspect of the result is that location-only predictions perform better when compared to location-and-delay predictions. The reason is that for location-and-delay predictions, the inaccuracy in the delay prediction adds to the inaccuracy of the location prediction. However, the discussed prediction scheme yields still accurate values within 1 meter in more than 90% of the cases.

Fifth, the routing approach and its location-and-delay predictions heavily depend on the *updates*. Figure 5.5 shows the accuracy of the location-and-delay predictions obtained for various constant values of the inter-update intervals for type-1 updates. As expected, when the update interval increases, the inaccuracy in the preditions increases. The curves merge for prediction errors of larger than 10 m. This is because these errors are caused by sudden change in the motion pattern. The changes occur at a much smaller time-scale than the inter-update intervals for the type-1 update. These *residual errors* occur before the type-2 update indicating the changed pattern of motion can propagate to the correspondent node.

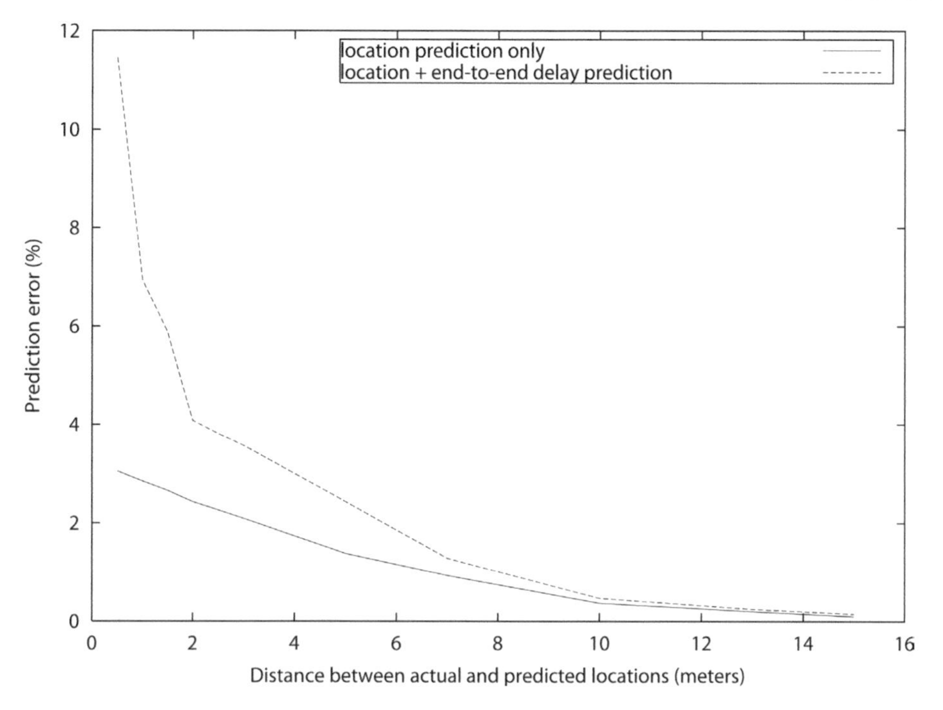

FIGURE 5.4: Accuracy of location + delay predictions vs location-only predictions [54].

Sixth, we want to confirm the *validity of having type-2 updates* and their impact on *location-only prediction*. Note that type-2 update is generated if the node's predicted location deviates from an actual/expected location by more than δ meters. Figure 5.6 shows the *location-only prediction error* represented by the distance between the actual and predicted location for various values δ. The location calculation happened at regular intervals of 40ms over the entire simulation run of 150 s. The prediction error metric here is the cumulative number of errors rather than the pecentage of errors so that one can show how the choice of δ affects the distribution of errors over various error distances. The results show that the number of errors is uniformly distributed for error distances less than δ for both curves. The number of errors for error distances greater than δ is considerably fewer because our type-2 updates filter out most of these errors. This stresses the value of the type-2 updates. Some residual errors, however, remain because of the delay in detecting the deviation of δ from the expected location and the delay in propagating the type-2 update.

Seventh, the question is what is the efficiency of the QoS routing algorithm in terms of *success rate of packets* delivered to the destination, and therefore providing *support for data accessibility service in application/middleware layers*. This depends on the *group environment*. Under the consideration of *group size* of 40 nodes, spread over *region 1100 m × 1100 m*, with nodes *velocity* between 0 and 15

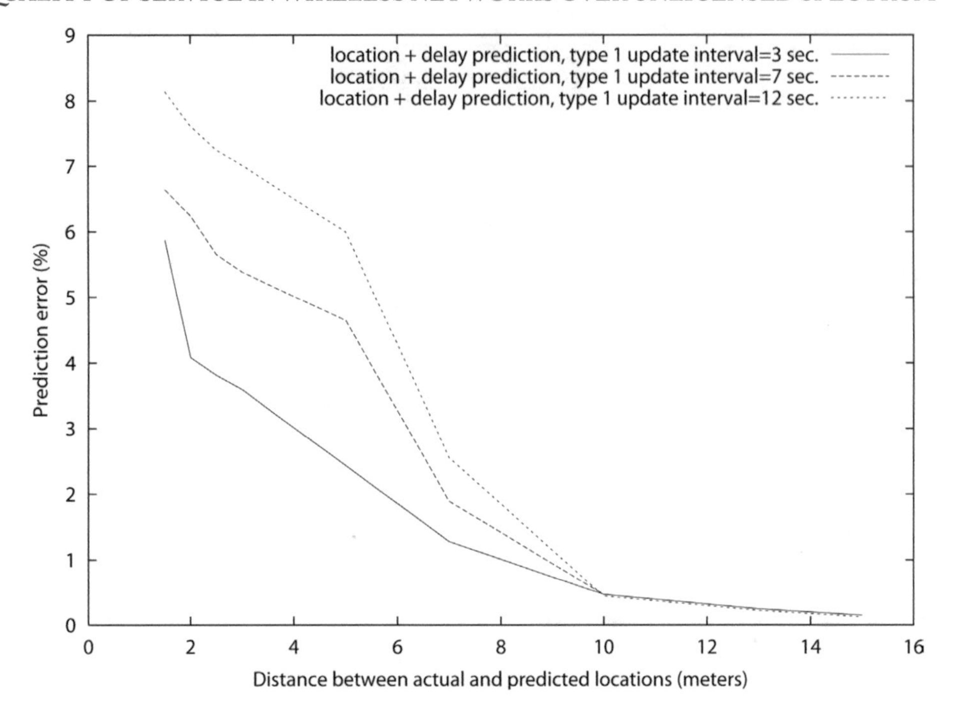

FIGURE 5.5: Accuracy of location + location prediction for various type-1 update intervals [54].

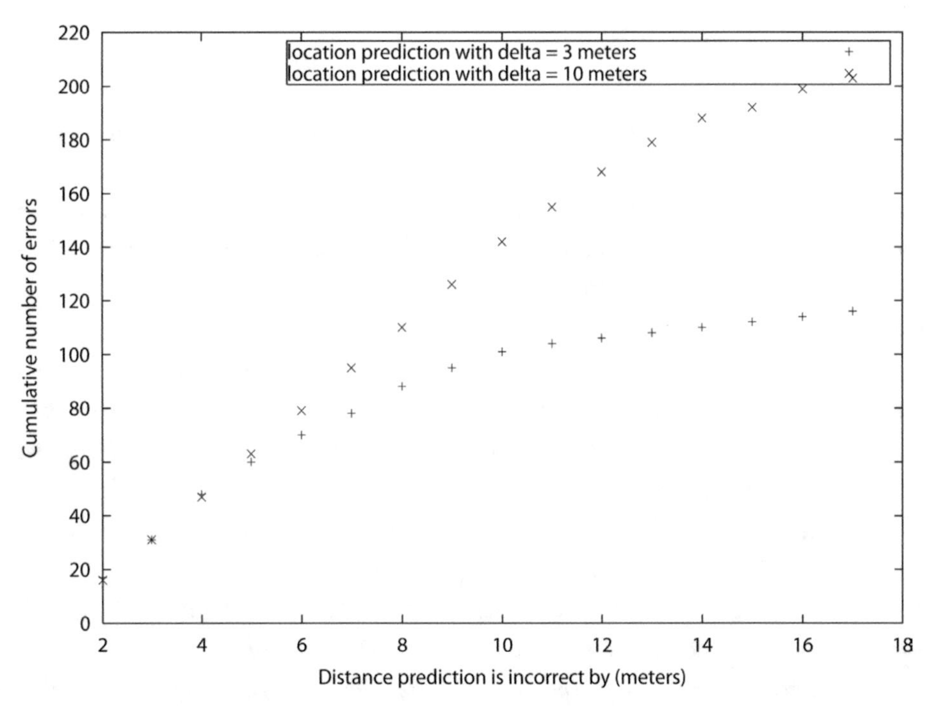

FIGURE 5.6: Efficiency of type-2 updates in location prediction for various error distances δ [54].

m/s, *background traffic* of 10 TCP connections between various nodes, type-1 *update interval* 3 s and with $\delta = 3$ m, few results emerge:

- *Route recomputation* had to be performed on average every 0.53 s;
- *Success ratio* of packet delivery was 69.10%;
- *Average throughput* measured in terms of packets per second to the destination was 141.91 pkt/s.

These performance numbers indicate that (a) with *additional information* such as location and delay, (b) with certain level of *regular mobility patterns* of mobile nodes (e.g., phones) in a group, and (c) *cross-layer services* to enable QoS adaptation and data replication in case of group split prediction (based on location predictions and type-2 updates), one can provide a *good data delivery* (but not great) with statistical QoS requirements and hence some level of accessibility success ratio among nodes in mobile wireless ad hoc scenario. Note that mobile wireless and ad hoc scenarios are difficult environments, but as new sensors and heterogeneous networks technologies emerge on mobile nodes (e.g., current mobile phones support multiple network technologies such as cellular, WiFi, Bluetooth and several sensors such as accelerometer, cameras, compas), it will be interesting to see how these *additional sensors and multinetworking* will assist in improving the data routing, delivery and data accessibility.

5.3 FAULT-TOLERANT ROUTING

Many existing designs of mobile wireless ad hoc networks are based on the assumption of cooperative and well-behaved, non-adversarial environments, where each node in the network cooperates in efficient routing and forwarding packets to the next best hop and in optimizing the end-to-end route according to a QoS performance metric. When *misbehaving nodes* (malicious or selfish) exist in the network, the QoS performance of current routing protocols *degrades* significantly. For example, if a misbehaving node, participating in the routing operation to route packets from node *a* to node *b*, *drops data packets* constantly (e.g., due to black hole attack), then a large number of packets will be lost. Misbehaving nodes exist for the following reasons:

(a) mobile nodes *lack adequate physical protection*, making them prone to be captured, compromised, and become misbehaving nodes,

(b) wireless ad hoc networks have *open medium and dynamic topology*, making them vulnerable to malicious external attacks,

(c) mobile hosts are severely *resource-constrained* computing devices, hence overloaded hosts lack the CPU cycles, buffer space or available bandwidth to forward packets. Selfish nodes are unwilling to spend their precious resources cooperatively for operations such as *packet forwarding* that do not directly benefit them.

To address the problem of misbehaving nodes, researchers have been looking for *fault-free network* utilizing two traditional design philosophies:

- The first one is *prevention* by *placing extra protection* on the routers, set certain rules to enforce cooperation in routing, etc. (e.g., [49, 50, 51]). But, since these mobile ad hoc networks are open and distributed in nature, maintaining prevention is extremely **difficult** and **expensive**.
- The second philosophy is *detection* (e.g., [64, 65, 66]). Here, nodes with *abnormal behaviors* are *detected* and *marked,* thus avoided by well-behaved nodes during packet routing. However, without a *centralized monitoring point* and network administrative authority, it becomes extremely *difficult* and *expensive* to effectively detect faulty nodes.

In summary, the goal of both designs is to acquire a *fault-free network* by preventing its nodes from selfinishess or malicious attacks, or purging nodes with such behaviors. This goal is *too expensive* to reach (if not at all impossible) in mobile wireless ad hoc network, where environment is open and distributed, and consists of autonomous nodes.

Third option is to deploy *fault-tolerant routing* which tolerates selfish and misbehaving nodes, and provides packet routing service with good QoS performance such as high delivery ratio and low overhead. There are some approaches that work with selfish nodes by *providing credits* to encourage packet forwarding via *incentives*, and then provide *cheat-proof system* for such credit exchange [e.g., 67, 68, 69].

We present a *fault-tolerant source routing algorithm*, called *BFTR (Best Effort Fault-Tolerant Routing)* [47], which instead of judging whether a path is good or bad, i.e., whether it contains any misbehaving node, it evaluates the *routing feasibility* of a path by its *end-to-end performance* (e.g., packet delivery ratio and delay). By continuosly monitoring the routing performance, fault-tolerant routing can then dynamically route packets via the most *feasible path.* We want to present the understanding of the design philosophy of QoS routing in challenging wireless environments with misbehaving nodes, and how to go about designing a routing algorithm and protocol for these environments. The presented routing algorithm [47] has some of the characteristics, we discussed in Section 5.1: (a) it runs over *mobile ad hoc networks*, (b) it cares about QoS performance metrics such as *packet delivery ratio* and *end-to-end delay*, (c) it provides *fault-tolerance* in adversarial envitonment, (d) it is a *source routing* approach, and (e) it comes up with a *point-to-point* network route.

In Section 5.3.1, we will discuss design choices as well as the algorithm overview, in Section 5.3.2, we present the routing protocol, and in Section 5.3.3, we discuss practical considerations. We will use the notation and definition in Table 5.2.

TABLE 5.2: Notations and definitions for Section 5.3.

NOTATION	DEFINITION
π	Path under testing
π^*	Feasible path
Ω	Path set after route discovery procedure
n	Number of random experiments on path π by sending n packets on it
$Y(\pi)$	Random variable representing the count if a packet, sent over path π, arrived at desired destination; it has value 1 if a packet arrived, it has value 0 if a packect did not arrive at desired destination
$p(\pi^*) = \Pr(Y(\pi^*) = 1)$	Probability that a packet is delivered successfully to desired destination
α	Significant level of test
$W(\pi,n) = \sum_{i=1}^{n} Y_i(\pi)$	Aggregated testing result from experimenting with n sent packets over the path π
W	Number of packets successfully passing through π and arriving at destination
p_0	Probability threshold that π is a feasible path, i.e., $p(\pi^*) \geq p_0$
w^*	Threshold number of arriving packets to determine if path π is infeasible
n_0	Window size
$d(\pi)$	Path delay
D	Desired end-to-end delay of a path
S; $\mathcal{D}$	Source node; destination node
H_0	Hypothesis that path π is feasible
H_1	Hypothesis that path π is infeasible
$b(w;n,p_0)$	Binomial distribution $b(w;n,p_0) = \binom{n}{w} \cdot p_0^w (1-p_0)^{n-w}$
$f(\alpha,p_0,n)$	Solution to the equation $\sum_{i=1}^{w^*} b(i;n,p_0) = \alpha$

5.3.1 Routing Algorithm Overview

The adversarial environments will include good and bad nodes enabling selection of good and bad paths as shown in Figure 5.7. We assume that a *good node* exhibits always the same behavior: *delivering packets correctly with high delivery ratio*. A *bad node* may exhibit *varying behaviors* such as packet dropping, tampering, misrouting, and others. Consequently, a *good path*, consisting of purely good nodes, always exhibits the same behavior pattern such as high delivery ratio from the end-to-end point of view. Any path which *deviates* from such a pattern is considered a *bad path*. For example, let us consider examples in Figure 5.7. If ***S*** is the source and ***D*** is the destination, then in Figure 5.7a *no good path* exists, since all routes from ***S*** to ***D*** must go through misbehaving nodes. In this case, *no routing protocol* can resolve this problem and find a good route. In Figure 5.7(b), both good and bad paths exist from ***S*** to ***D*** in the network. In this case, it is possible to deliver packets correctly.

We make also other *design choices/assumptions* on the *security* support of the network:

(1) Both the source and destination nodes of the connections that are considered in our design are *well-behaved* nodes. It means, routing service is not provided to misbehaved sources and destinations.
(2) There is *a priori trust relationship* between the source and the destination so that they can authenticate each other during data communication. In this way, their messages (data and acknowledgement) cannot be forged by anyone in the middle. There are various approaches to instantiate such *trust relationships* [e.g., 70, 71, 72].

Next, we present the design of the fault-tolerant *BFTR routing algorithm*. Routing algorithms consist usually of three main steps: (1) *route discovery*, (2) *route selection*, and (3) *route maintenance*.

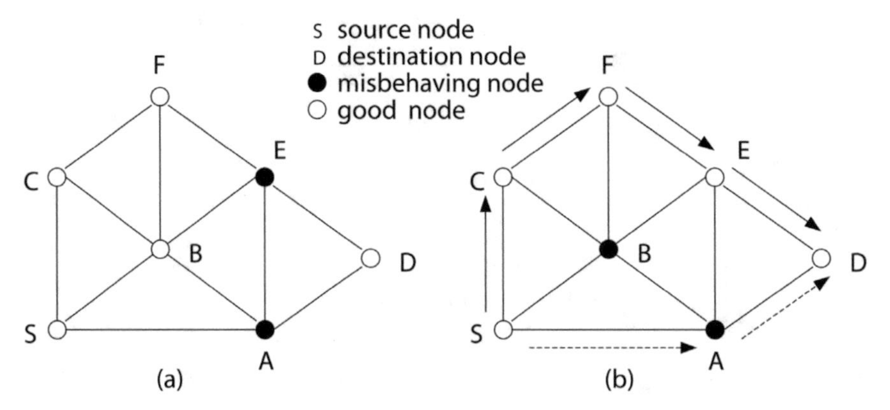

FIGURE 5.7: Path scenarios: (a) No good path exists between source ***S*** and destination ***D***; (b) The shortest path ***S***–A–***D*** from ***S*** to ***D*** is bad; ***S***–C–F–E–***D*** is a good path [47].

BFTR follows these steps. In the route discovery, BFTR uses the *DSR (Dynamic Source Routing)* [48] flooding concept to retrieve a set of paths Ω between source and destination nodes, whenever necessary. Then BFTR selects the shortest path π from this set Ω to send packets. If within a certain *observation window*, this path is identified as *feasible*, the algorithm continues to send packet along it. The window keeps shifting during the packet sending to observe the most recent behavior of the same path. Anytime it is identified as an *infeasible path*, the algorithm *discards* the path/route and selects the next shortest path to repeat the same procedure. The summary of the BFTR algorithm is as follows:

1. *Route discovery*: BFTR retrieves a set of paths Ω.
2. *Route selection*: Upon retrieving a set of paths via route discovery, BFTR selects a path π to route packets along the shortest path from the path set Ω. Using the *test heuristic* over a certain observation window, BFTR identifies and determines whether this path is *infeasible* and should be *rejected*.
3. *Route maintenance*: If the current path is rejected, it chooses the second shortest path to repeat this procedure. If all paths in Ω have been rejected, a new route discovery procedure is triggered to retrieve a new set of paths, i.e., to renew Ω. The testing procedure then restarts every time Ω is renewed.

5.3.2 Finding Feasible Paths

The major issue within the BFTR algorithm is to consider the problem of *finding feasible paths*. We present a *testing heuristic* approach to maximally maintain routing end-to-end QoS performance. The main criterion of selecting a path is its *feasibility* with respect to requested QoS performance, e.g., whether a path is able to transmit packets correctly with *high delivery ratio*. Any path that follows this pattern can be regarded as a *feasible path* and a path which deviates from this pattern is an *infeasible path*. On the basis of this idea, *data packets transmission* is used as a tool to measure the end-to-end QoS performance. By comparing the *measured path performance* and *a priori feasible path model*, the testing heuristic determines the feasibility of a path. The presented testing heuristic is based on end-to-end performance, thus requiring no additional support from the intermediate nodes, which might be misbehaving. Also, from an end-to-end point of view, the BFTR algorithm can provide a unified solution for many types of misbehavior. Before we discuss the *testing procedure*, to determine whether a path π is feasible, we introduce "a priori feasible path model."

Feasible path model. Given a path π, we define a random variable $Y(\pi)$ as follows:

$$Y(\pi) = \begin{cases} 1 & \textit{if } \pi \textit{ correctly delivers the observed data with delay } d(\pi) \leq D; \\ 0 & \textit{otherwise.} \end{cases}$$

A path π^* is a *feasible path*, if the probability $p(\pi^*)$ that a packet is delivered successfully (i.e., $\Pr(Y(\pi^*)=1)$ satisfies $p(\pi^*) \geq p_0$. It also means, for a feasible path π^*, $Y(\pi^*)$ follows the *Bernoulli distribution*, i.e., $\Pr(Y(\pi^*)=1) = 1 - \Pr(Y(\pi^*)=0) = p(\pi^*)$. The probability $p_0 \in [0,1]$ can be understood as the *expected packet delivery ratio*, and it is determined from field measurements. It means p_0 may vary, depending on the network load and the wireless channel quality.

Determination process of a feasible path. Given the model of a feasible path, we now need to *determine whether a path π is feasible.* Intuitively, if the packet delivery behavior over the path π follows the feasible path model, then this path is likely to be a feasible path, otherwise, it is more likely to be an infeasible one. We present a procedure design that will *identify infeasible path* using the method of *hypothesis testing*. This testing procedure is simple, and with this hypothesis testing we can quickly determine if a path is infeasible.

First, we establish the *one-tailed test* (testing only one direction of our interest) with the null hypothesis $\boldsymbol{H_0}$ and the alternative hypothesis $\boldsymbol{H_1}$ as follows:

$\boldsymbol{H_0}$: path π is a *feasible path*, i.e., $p(\pi) = p_0$.
$\boldsymbol{H_1}$: path π is an *unfeasible path*, i.e., $p(\pi) < p_0$.

Note that for the feasible path π^* definition, we have $p(\pi^*) \geq p_0$. Yet, in the one-tailed test, "=" is used instead of "≥" conventionally. The reason is that we are interested to have the hypothesis testing only in one direction, i.e., if $p(\pi) < p_0$ (path is infeasible) or not.

Within the hypothesis testing, we conduct n random experiments on path π by sending n packets on it (sending one packet counts as one experiment). The outcome of the testing will be sequence of 0 (packet was not correctly delivered) or 1 (packet was correctly delivered within the deadline) random samples $Y_1(\pi), Y_2(\pi), \ldots, Y_n(\pi)$. If we aggregate the testing results from the testing of n packets, i.e., $W(\pi,n) = \sum_{i=1}^{n} Y_i(\pi)$, then we get w (outcome of $W(\pi,n)$) packets out of n packets pass through the path π to the destination. If the number of arrived packets w is below a certain threshold w^*, then the testing procedure rejects the path π and declares the path to be infeasible.

Formally, $\boldsymbol{w}$ will determine whether or not to reject H_0, and we seek an *approach* how to determine whether or not to reject H_0 and what w^* should be. We take the following approach:

(1) *We determine $\Pr\{W(\pi,n) = w \mid H_0$ is true$\}$:* If H_0 is true, then $W(\pi, n)$ is a *binomial random variable* with distribution:

$$\Pr\{W(\pi,n) = w \mid H_0 \text{ is true}\} = b(w;n,p_0) = \binom{n}{w} \cdot p_0^w (1-p_0)^{n-w} \tag{5.1}$$

The reason for $W(\pi,n)$ being a binomial random variable comes from the statistics that states that since $Y_1(\pi), Y_2(\pi), \ldots, Y_n(\pi)$ are all independent, identically distributed (i.i.d.) random variables and these variables are Bernoulli distributed with success probability p_0, therefore: $W(\pi,n) = \sum_{i=1}^{n} Y_i(\pi) \sim \text{Binomial}(n,p_0)$.

(2) *We introduce a significant level of test*: Intuitively, if the probability $\Pr(W(\pi,n) \le w \mid H_0$ is true} is small, then H_0 is likely to be false. Let us introduce α as a threshold of rejecting H_0, and α is also the probability of rejecting a true hypothesis. Hence, α represents the significant level of the test. Then we get

$$\Pr(W(\pi,n) \le w \mid H_0 \text{ is true}\} < \alpha, \qquad i.e., \qquad \sum_{i=0}^{w} b(i;n,p_0) < \alpha \qquad (5.2)$$

(3) *We construct a critical region of the outcomes*, where H_0 is rejected: All outcome values w, satisfying Equation 5.2, construct a critical region R. If the outcome of $W(\pi,n)$ falls into this region, i.e., $w \in R$, then we reject H_0 in favor of H_1, which means that path π is regarded as an infeasible path. Moreover, the probability that such rejection is wrong is less than α. That is, we reject H_0 at the significant level α. Note also that α takes very small values (e.g., 0.05 or 0.0.1) in the hypothesis testing.

(4) *We determine threshold w^**: Given the feasibility path model with expected packet delivery ratio p_0, and significant level of the test α, we can then determine whether the measurement result from ***n*** samples is sufficient to reject H_0 and calculate the threshold w^* from $\sum_{i=0}^{w^*} b(i;n,p_0) = \alpha$. This equation can be transformed into a function $f(\alpha, p_0, n) = w^*$; i.e., the function $f(\alpha, p_0, n)$ represents the solution to $\sum_{i=0}^{w^*} b(i;n,p_0) = \alpha$. For each α and p_0, $f(\alpha, p_0, n)$ can compute the threshold w^* for each set of n random experiments. If w, the outcome of $W(\pi,n)$ based on n random experiments, is less than the computed value of w^*, i.e., $w < w^*$, then the path is rejected as an *infeasible* one. Otherwise, we keep on sending packets along this path.

Example: Let us illustrate some of the statistical hypothesis testing concepts on an example. Let us consider the binomial distribution of arrived packets over path π, $W(\pi,n)$, as shown in Figure 5.8a with $n = 40$ sent packets, and the expected packet delivery ratio $p_0 = 0.9$ (i.e., probability that π correctly delivers the observed data packet with delay $d(\pi) \le D$). The shadowed region shows that the probability that the outcome of $W(\pi,n)$ is less than or equal 32 (which is $\sum_{i=0}^{32} b(i;40,0.9)$) is less than 0.05 (the value of α). The corresponding values of w construct the critical region R [0,32]. Thus, if the outcome of $W(\pi,n) \in [0,32]$, then the path π is *rejected* as an *infeasible* path at the significant level 0.05. This means that the probability that the path π is feasible is less than 0.05. Figure 5.8b shows the function $f(\alpha, p_0, n)$ for different α's. We can see that the threshold w^* will be 33 packets for $n = 40$ and probability $p_0 = 0.9$.

Discussion. This testing procedure indicates whether path is *infeasible*, but cannot confirm whether a path is feasible because of the one-tailed hypothesis testing procedure design. In our

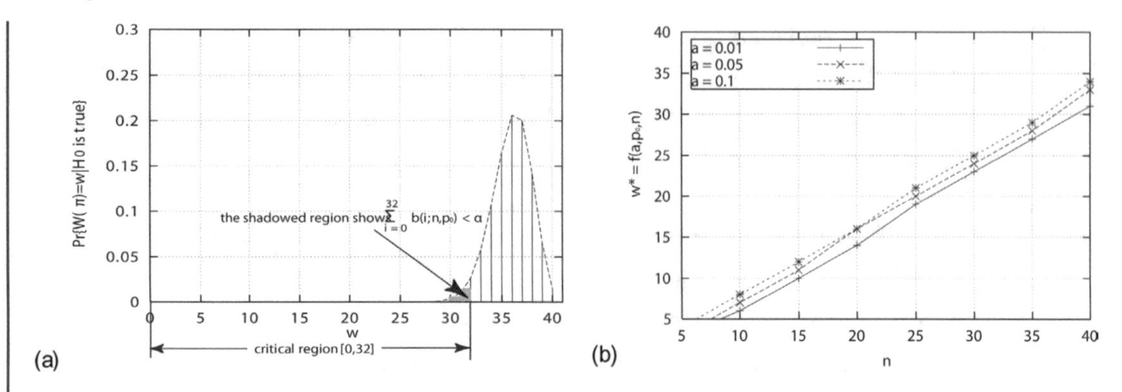

FIGURE 5.8: Example with $n = 40, p_0 = 0.9, \alpha = 0.05$: (a) Critical region R and (b) $w^* = f(\alpha, p_0, n)$ [47].

routing algorithm it means that we keep sending packets along one path until the testing procedure indicates it is an infeasible path. As n (number of packets sent) increases, the probability that an infeasible path is not rejected monotonically decreases. On the other hand, as n gets larger, the procedure becomes more insensitive to the path with inconsistent misbehaviors. For example, a well-behaved node on path π may suddently start to misbehave after successfully delivering a large number of packets. In this case, the testing procedure will respond slowly as it needs enough packet delivery failures to reject π. To balance this trade-off, we introduce an *observation window*, whose size n_0 is the maximum number of packets to look back. If more than n_0 packets have been sent along π, then the algorithm only uses the most recent n_0 packets deliveries to determine whether to reject π.

5.3.3 Routing Protocol Design

The BFTR algorithm needs to be realized as a *protocol* among the mobile ad hoc nodes and in this subsection we will discuss the *BFTR protocol design*. We will design the BFTR algorithm as an *on-demand source routing protocol*. Every data packet in BFTR has a source path in its header, which consists of the node addresses through which the packet should be forwarded to reach the destination. We use an on-demand routing protocol design because *on-demand routing* protocols have been demonstrated to perform better than proactive routing protocols with significantly lower overhead under many scenarios. The protocol design will follow the three main steps of the BFTR algorithm, the route discovery, route selection and route maintenance.

Route discovery. The route discovery of BFTR is similar to DSR [48]; hence, we briefly describe the DSR's route discovery protocol.

DSR route discovery. In DSR, each mobile node maintains a route cache in which it keeps the source routes that it has learned. When a node sends a packet, it first checks its route cache for a route to the destination. If a route is found, then packets are transmitted along this path; otherwise, the source node initiates route discovery to find a new route. In route discovery, a route request packet (RREQ) is flooded, specifying the target of that discovery and a unique identifier. In the fooding procedure, each intermediate node that receives RREQ appends its own node address to a list in RREQ and rebroadcasts RREQ, if it has not recently seen this request identifier from the source; otherwise, it discards this RREQ. When RREQ reaches its destination, the destination sends a route reply packet (RREP) back to the source including a copy of the list of addresses from RREQ either along a known route or the reverse path of RREQ. When RREP reaches the source, it is added as a new route in the route cache.

BFTR route discovery. BFTR is similar to DSR in route discovery, except that

(1) BFTR requires the RREP packet to be sent *along the reverse path* of the RREQ packet;
(2) BFTR requires the destination to *send multiple replies*, so that the source can have *multiple paths* between the source and the destination;
(3) RREP packets are signed with the *shared secret key* between the source and the destination to prevent them from *fabrication* and *replay attacks.*

Route selection. Given the set of paths Ω acquired through route discovery, the source node makes its local decision in path selection using the BFRT hypothesis testing procedure (test heuristic). The following issues are addressed in the protocol design:

a. *w evaluation*: The source needs to know the number of received packets $\boldsymbol{w}$ along the path under testing. To achieve this, the destination needs to send a *feedback* to the source to inform it about the value of w. The feedback can be implemented by piggybacking onto TCP acknowledgment (ACK) packets or through additional ACK packets in the network layer. The ACK packets are signed using the shared secret key with sequence numbers to protect them from fabrication and replay attack. In this way, only the authenticated ACK packets from the destination will be accepted, thus ensuring the correct w evaluation.
b. *Packet delay*: In BFTR algorithm, packets are regarded as successfully delivered, if they are received correctly with delay less or equal to $\boldsymbol{D}$, i.e., $\mathrm{d}(\pi) \leq D$. Thus, for each packet sent, the source node sets a timer with value $2 \cdot D$. If the timer expires and no ACK is received, then the packet is regarded as lost. In this way, the round trip delay of the packet delivery is implicitly used as end-to-end performance to determine feasible paths. Actually, large delay in packet delivery does not necessarily mean that the path in use is infeasible. Yet, it at least indicates bad network condition along the path such as congestion or long route. For this reason, BFTR prefers paths with small bounded delay ($\boldsymbol{D}$). Setting the appropriate value of $\boldsymbol{D}$ is an important issues. Too small $\boldsymbol{D}$ will wrongly reject feasible paths; too large

D will accept infeasible paths (e.g., misbehaving nodes delay the packets for some time), resulting in large end-to-end latency. One possible guiding value can be *two times the expected round trip* time of all the paths as ***D***. Round trip time information can be acquired through route discovery.

Route maintenance. BFTR maintains routes the same way DSR does. In the DSR route maitenance, if a node detects a link failure, it returns a *route error packet* (RERROR) to the source, identifying the broken link from itself to the next node. The source node then removes the identified broken link from its route cache. For subsequent packets to this destination, the source may use any other route to that destination in its route cache, or it may attempt a new route discovery after a certain back-off time interval. In BFTR, if a route failure report is received, the protocol discards the current routing path and proceeds with the next shortest path in the route cache. Moreover, if all paths in the current route cache have been rejected, BFTR will initiate new route discovery just as what DSR does, attempting to discover mode paths. Note that BFTR does not distinguish between route failure due to the mobility and test failure caused by misbehaving nodes, as in either case the path is infeasible. Also to control the overhead of possible repeated route discover, BFTR adopts the same approach as DSR, which spaces out the route request with back-off intervals.

5.3.4 Practical Issues

Every protocol design needs to be considered under different more practical situations to evaluate its suitability for the given set of requirements and environments, where it needs to operate. We analyze the BFTR in terms of the routing operations under various node misbehaviors as well as its operations in comparison with other protocols and within its own parameter design space.

First, BFTR can *ensure the correct routing* operation under various node misbehaviors. We show few misbehaviors and BFTR's performance in the presence of these misbehaviors.

I. *Dropping*: If data packets are *dropped* by misbehaving nodes, the testing heuristics can effectively identify the paths so that the paths in use will give high packet delivery ratio. If either RREQ or RREP packets are dropped, the path on which misbehaving nodes are located will not be discovered. Thus, the routing performance will not be affected, since BFTR is unaware of this bad path. If ACK packets are dropped, BFTR will have a low measurement on the packet delivery ratio of this path (w/n), which lowers the probability that we identify this path as a feasible one.

II. *Corruption*: If packets are *corrupted*, the source and destination nodes can detect the corruption through error detection code and discard the packets. Therefore, packet corruption can be treated the same as packet dropping.

III. *Misrouting*: When data packets are routed to a *wrong next hop*, the node at the next hop will find that it is not on the source route, thus drop the packet. If the node at the next hop is also a misbehaving node, and continues to forward the packets, two scenarios may happen from the destination point of view: (a) packets do not arrive at the destination within the required delay constraint; (b) packets reach the destination and satisfy the delay constraint. In the first case, delayed packets are regarded as *lost* packets. BFTR can successfully reject this path. As a special example, if several misbehaving nodes collude to forward data packets along a very long path, which results in long delay in packet delivery, then timeout will be triggered at the source. BFTR can identify this type of attack. If the second scenario happens, it means that even the misrouted path may consist of misbehaving nodes, it can still deliver the packets in a preferred way and will be used in BFTR. Similar situation happens when ACK packets are misrouted.

IV. *Tampering*: If the source route is changed, then the *tampered path* is tested. From an end-to-end point of view, two scenarios may happen: (a) the path in use has small w. In this case, the source will discard this path; (b) the tampered path is able to deliver packets with a high rate. Then this path will be used for packet delivery, even if there may exist bad nodes on this path. The packet delivery ratio of BFTR will not be greatly affected by mis-selecting such paths. One problem with BFTR is that if many misbehaving nodes collude to claim they form a short path, BFTR cannot identify this scenario, and may choose this sub-optimal path. If the routing information in ACK packets is changed, ACK packets will be mis-routed, resulting in the same situation as in misrouting. Other changes except route change in ACK packets or RREP packets will not affect the performance of BFTR. Tampering the signed fields of ACK or RREP packets will corrupt packets, resulting in the same situation as in corruption.

V. *Delaying*: Delayed packets will be viewed as either packets dropped or packets delivered at the end-hosts, depending on how long the packets are delayed. If the packets are *delayed* for a time larger than D, the source node can correctly identify the bad path; otherwise the source may mis-identify the bad paths as a good one. However, the routing performance will not be greatly affected by such mistakes, since only a small amount of end-to-end latency can be introduced by these mistakes.

VI. *Fabrication and replay attack*: The misbehaving nodes cannot successfully *fabricate* or *replay* RREP or ACK packets, as they are signed with sequence numbers.

VII. *Faked route failure report*: BFTR will abandon the current routing path once a route failure report is received, regardless its truthfulness. The reason is that the path is infeasible for routing in either way. It is either broken, or contains a misbehaving node which refuses to relay packets.

Second, BFTR can maximally maintain the routing performance in the presence of node misbehaviors via *end-to-end performance measurements*, provided by the ACK packets. As long as ACK packets indicate the successful data packets receipt (guaranteed by the destination's signature), BFTR can make a preferable route selection decision. However, BFTR is a *best-effort protocol* since it does not guarantee routing performance. The probability of packet delivery failure will inevitably increase when more misbehaving nodes are present, i.e., success rate will decrease with increasing number of misbehaving nodes as shown in Figure 5.9. Such a condition may also happen when the network becomes heavily loaded. From the view of the BFTR protocol, these two cases are indistinguishable. However, both of them indicate that the network is infeasible for routing: no protocol will work well in these conditions.

Third, the BFTR *packet delivery ratio is higher* than DSR [48] and watchdog [73] in the same adversarial environment, and the packet delivery ratio of BFTR drops more slowly than DSR as the number of misbehaving nodes increases, as shown in Figure 5.10a. The reason is that BFTR has a much higher success rate in using the feasible path than DSR as shown in Figure 5.9. Using a feasible path to deliver packets obviously increases the packet delivery ratio of BFTR. The reason that BFTR has a higher packet delivery ratio than watchdog, especially in case of increasing misbehaving nodes, is that the increase leads to higher probability of successful collusions which cannot be detected by watchdog.

Fourth, the design of BFTR *incures an overhead* by the destination's feedback. This overhead can be saved by cooperating with transport-layer protocol such as TCP. The current design

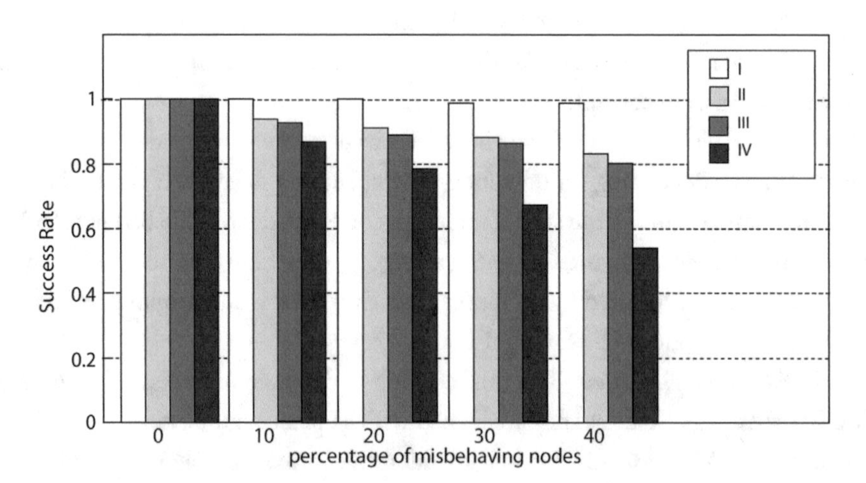

FIGURE 5.9: I Percentage of scenarios where network has a good path; II Percentage of scenarios where flooding discovers a feasible path; III Percentage of scenarios where BFTR uses feasible path; IV Percentage of scenarios where DSR uses a feasible path (results from ns-2 simulations, using 802.11 radio/MAC, random waypoint mobility, node speed between 0 and 20 m/s, sending rate 2 kbps) [47].

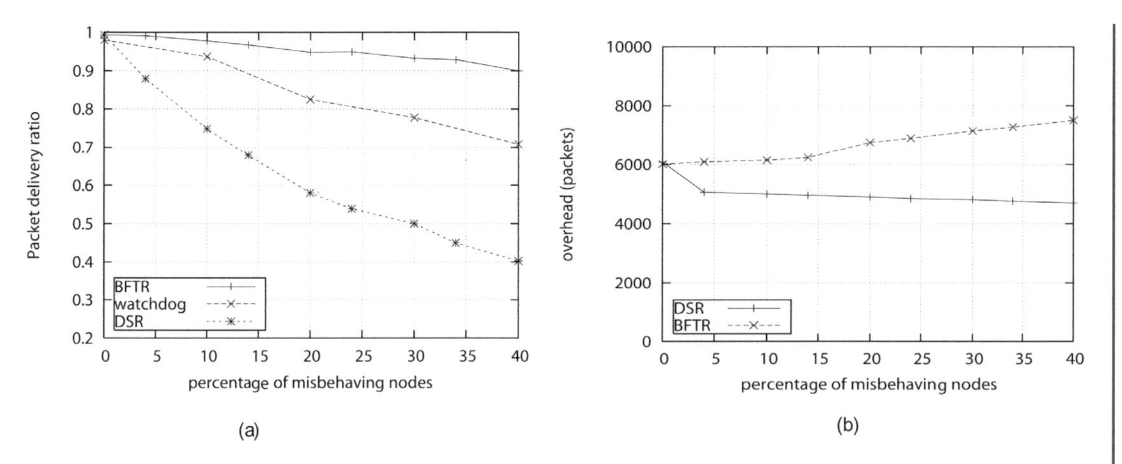

FIGURE 5.10: (a) Packet delivery ratio vs percentage of misbehaving nodes; (b) protocol overhead vs percentage of misbehaving nodes [47].

of BFTR has higher overhead than DSR as shown in Figure 5.6(b) especially when the number of misbehaving nodes increases. This is because more route discovery procedures may be triggered by BFTR as a result of absence of feasible paths in the initial rounds of testing. DSR's overhead slightly decreases with larger number of misbehaving nodes. This is because in DSR a large number of packets can be lost due to bad paths.

Fifth, during the design of the BFTR algorithm and protocol, we stressed the importance of several design parameters. One of the design parameters was the observation window size n_0 which determines at how many packets does the testing procedure look at to determine if a path is

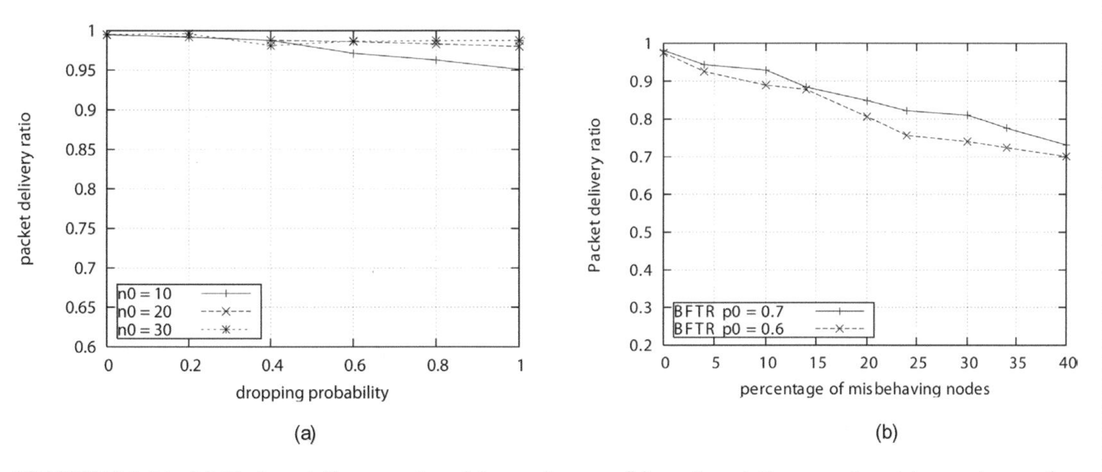

FIGURE 5.11: (a) Packet delivery ratio with varying n_0; (b) packet delivery ratio with varying p_0 (p_0 simulation with dropping probability 0.6) [47].

infeasible. Figure 5.11a shows the *impact of the window size* on the packet delivery ratio under dropping packet attack (with different dropping probabilities).

Sixth, another important parameter in the BFTR design is the probability p_0 that a path π is a feasible path. The effect of p_0 is shown in Figure 5.11(b). The results show that underestimating the parameter p_0 will lead to *imprecise feasible path testing results* and effect the performance of BFTR, especially when the misbehaving nodes only drop part of the packets. On the other hand, overestimating p_0 may make the testing algorithm yield no feasible path, result in repeated route discovery and lead to higher overhead. The results indicate that choosing p_0 from the range 0.9–0.95 gives the best performance of BFTR.

Overall, BFTR offers a *unified solution* to find routes. The route selection process is driven by QoS performance, as well as it can coexists with various types of misbehaviors in an *adversarial environment.*

5.4 ENERGY-EFFICIENT DISTRIBUTED ROUTING

One of the important issues in wireless network is the *energy constraint* and with it connected *power consumption* in wireless devices. The reason is that wireless nodes carry limited and irreplaceable energy supply. Moreover, radio communication on mobile devices (e.g., phones) consumes a large fraction of energy supply. Such constraints pose critical demands on the design of *energy-efficient routing algorithms* in mobile devices during packet transmission.

The problem of designing energy-efficient routing algorithms has been extensively studied in both general *multi-hop wireless networks* [e.g., 112, 113, 114, 115] and the particular backdrop of *sensor networks* [e.g., 116, 117]. We will briefly discuss *two aspects* of energy-based routing that one needs to understand when choosing a solution for given wireless networks.

Minimum energy routing problem presents a "user optimization" problem because it routes packets to optimize the energy usage of a single user (an end-to-end connection). To solve this problem, the typical approach [e.g., 118] is to use a *shortest path algorithm* in which the edge cost is the power consumed to transmit a packet between two nodes of this edge. Though effectively reducing the energy consumption rate, this approach can cause unbalanced consumption distribution, i.e., the nodes on the minimum-energy path are quickly drained of energy, causing network partition or malfunctioning.

Energy-based routing problem can be also looked at from the *overall network welfare* aspect to solve the *maximum network lifetime routing* problem. This problem tries to maximally prolong the duration in which the entire network properly functions. It presents a "system optimization" problem which is very different from "user optimization." To achieve "system optimization", global optimization and coordination is required, which poses significant challenges to the design of a routing algorithm. On the other hand, maximum lifetime routing addresses well the power consumption balance problem of the minimum energy routing.

In this section, we discuss the energy-based routing from the aspect of the *overall network welfare* and present a solution to *the maximum lifetime routing* problem. There exist numerous algorithms to this problem, some use *heuristic* solutions [e.g., 113, 116], some use *centralized approximation* solutions [e.g., 119], and some use *distributed* solutions [e.g., 112]. The presented approach will be a fully *distributed routing* algorithm [42] that maximizes the overall network lifetime.

In Section 5.4.1, we describe the overall network and traffic demand concepts under which we study the routing problem. Section 5.4.2 presents the energy and power consumption concepts. Formulation of the maximum network lifetime routing problem as an energy-constrained resource optimization problem will be discussed in Section 5.4.3. Section 5.4.4 gives the solution of the distributed routing algorithm to the maximum network lifetime routing problem. In Section 5.4.5, we discuss some of the practical considerations one needs to take into account when designing energy-efficient routing algorithms. In the routing framework we use definitions and notation shown in Table 5.3.

TABLE 5.3: Notations and definitions for Section 5.2.

NOTATIONS	DEFINITIONS
$N = \{1,\ldots,n\}$	Number of nodes in wireless network
$L = \{(i,k) \mid \text{wireless link from } i \text{ to } k\}$	Set of wireless links
$r_i(j)$; $\mathbf{r} = \{r_i(j) \mid i,j \in \text{N}\}$	End-to-end traffic rate generated at source node i and goes to destination j; end-to-end traffic set $\mathbf{r}$ is the set of all generated traffic rates
$t_i(j)$; $\mathbf{t} = \{t_i(j) \mid i,j \in \text{N}\}$	Total traffic node flow rate at node i destined for node j; node flow set $\mathbf{t}$ is the set of total flow rates
$\phi_{ik}(j)$; $\Phi = \{\phi_{ik}(j) \mid i,j,k \in \text{N}\}$	Routing fraction of traffic over link (i,k); routing set $\mathbf{\Phi}$ represents all routing fractions
E_i	Energy reserve at node i
p_i; $\mathbf{p} = \{p_i \mid i \in N\}$	Node i's power consumption rate (J/s); power consumption set $\mathbf{p}$ is the set of all power consumptions at all nodes in the network

(*continued*)

TABLE 5.3: (*continued*)

NOTATIONS	DEFINITIONS
p_i^r	Power consumption (J/bit) at node i when it receives one unit of data
p_{ik}^t	Power consumption (J/bit) at node i when it transmits one unit of data from i over link (i,k)
d_{ik}	Distance between antennas of node i and k
$T_i = E_i / p_i$;	Lifetime of node i
$T = \min\{T_i, i \in \mathrm{N}\}$	Lifetime of the wireless network with N nodes; The lifetime of the whole network is equal to the lifetime of the node i that dies first
T	Optimization problem to maximize network lifetime
U	Utility represents the degree of satisfaction of a user
$U_i(T_i)$	Utility function U_i at node i, depending on node i's lifetime T_i
$U = \sum_{i \in \mathrm{N}} U_i$	Aggregate utility U of all nodes within the network (N, L)
Commodity j	Name for "Flow routed to destination j"
Marginal utility of a good or service	Utility gained (or lost) from an increase (or decrease) in the consumption of that good or service
$\delta_{ik}(j) = U'_{ik} + \frac{\partial U}{\partial r_k(j)}$	Marginal utility $\delta_{ik}(j)$ of link (i,k) with respect to commodity j is the sum of the marginal utility U'_{ik} of link (i,k) and the marginal utility $\frac{\partial U}{\partial r_k(j)}$ of node k with respect to commodity j
$B_{i,\phi}(j)$	Set of blocked nodes k to avoid loops in routing, i.e., for which $\phi_{ik}(j) = 0$ and the algorithm is not permitted to increase $\phi_{ik}(j)$from 0

5.4.1 Network and Traffic Demand Models

The discussion of the distributed routing to maximize the network lifetime will consider a *multi-hop static wireless network* with $N = \{1, 2, ..., n\}$ nodes. Two nodes in the network, which are within the transmission range, communicate directly and form a *wireless link*. Let $L = \{(i,k) \mid \text{wireless link from } i \text{ to } k\}$ be the set of wireless links, where each link (i,k) has a weight d_{ik} representing the distance between the antennas of node i and k. Figure 5.12 shows an example network with 6 nodes $N = \{1,2,3,4,5,6\}$ and links $L = \{(2,1), (3,1), (1,4), (1,5), (5,6), (4,6)\}$ that we will use to show some of the concepts.

Let $r_i(j) \geq 0$ be the *end-to-end traffic rate* (in bits per second), generated at node i and destined for node j. For example, in Figure 5.12, the end-to-end traffic rate generated at node 2 and destined to node 4 is $r_2(4) = 2kbps$, the rate generated at node 3 and destined to node 6 is $r_3(6) = 3kbps$, and the rate generated at node 1 and destined to node 6 is $r_1(6) = 1k$.

Let $t_i(j)$ be the *total traffic rate* of node i destined for node j, including both $r_i(j)$ and traffic from other nodes that is routed through i to destination j. For example, in Figure 5.12, the total traffic rate from node 1 to node 6 is $t_1(6) = r_3(6) + r_1(6) = 3 + 1 = 4kbps$, total traffic rate from node 1 to node 4 is $t_1(4) = 2kbps$; the other total traffic rates are $t_2(4) = 2kbps$; $t_3(6) = 3kbps$; $t_4(6) = 1kbps$; and $t_5(6) = 3kbps$. Let us consider routing variable $\phi_{ik}(j)$ to be the *fraction of the total node flow rate* $t_i(j)$ routed over link (i,k). For example, in Figure 5.12, the fraction $\phi_{14}(6)$ of the total node flow rate $t_1(6) = 4kbps$ routed over link (1,4) from node 1 through node 4 to node 6 is $\phi_{14}(6) = 1/4$ and the fraction $\phi_{15}(6)$ of the total node rate $t_1(6) = 4kbps$ routed over link (1,5) from node 1 through node 5 to node 6 is $\phi_{15}(6) = 3/4$; the other fractions are $\phi_{21}(4) = \phi_{31}(6) = \phi_{14}(4) = \phi_{46}(6) = \phi_{56}(6) = 1$. This routing variable is important and defines a routing solution with

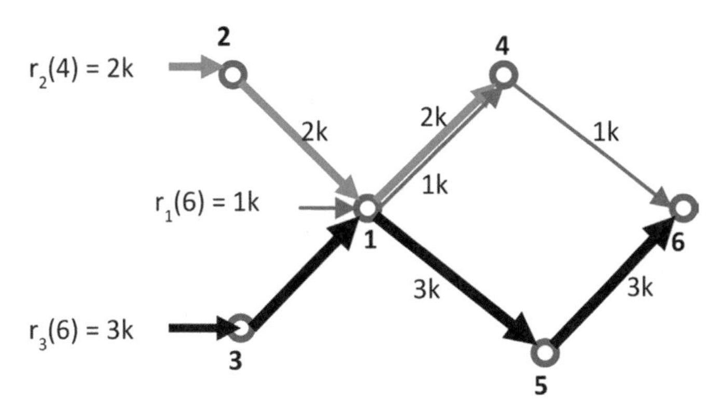

FIGURE 5.12: Example of five-node network.

- $\phi_{ik}(j) = 0, if (i,k) \notin$ L (no traffic can be routed through non-existing link);
- $\phi_{ik}(j) = 0$, if $i = j$ (traffic that has reached its destination is not sent back into the network);
- $\sum_{k \in N} \phi_{ik}(j) = 1, \forall i, j \in N$ (node i must route its entire node flow rate $t_i(j)$ through all outgoing links)

For example, in Figure 5.12, the node 1 routes the node flow rate $t_1(6)$ through both outgoing links (1,4) and (1,5) with $\phi_{14}(6) + \phi_{15}(6) = 1$; on the other hand, $\phi_{6k}(6) = 0$ and $\phi_{23}(6) = 0$.

One important constraint of traffic routing is the *flow conservation*, i.e., *the traffic into a node for a given destination is equal to the traffic out of it* for the same destination. For example, in Figure 5.12, the traffic rate flowing *into node 1* toward the destination node 6 is $r_1(6) = 1kbps$ and also $r_3(6) = 3kbps$ which is of total node incoming rate $t_1(6) = 4kbps$. The same total amount of traffic $t_1(6)$ flows out of the node 1, where ¼ fraction of the total traffic $t_1(6)$ routes over link (1,4) and ¾ fraction of the total traffic $t_1(6)$ routes over link (1,5). This flow conservation can be expressed more formally with Equation 5.3:

$$t_i(j) = r_i(j) + \sum_{l \in N} t_l(j) \phi_{li}(j), \forall i, j \in N \tag{5.3}$$

Using Figure 5.12 as an example, *the total incoming traffic rate* going into the node *1* (including local node 1 generated traffic) with final destination node 6 is $r_1(6) + t_2(6) \cdot \phi_{21}(6) + t_3(6) \cdot \phi_{31}(6) + t_4(6) \cdot \phi_{41}(6) + t_5(6) \cdot \phi_{51}(6) = 1 + 0*1 + 3*1 + 1*0 + 3*0 = 4k$; the total outgoing traffic rate from node 1 to final destination node 6 is $t_1(6) = 4kbps$, hence, the flow conservation is preserved.

5.4.2 Power Consumption Model

To model power consumption and energy in wireless network, we first make a careful distinction between *energy* and *power*. Power is the rate at which work is performed or at which energy is converted. It means $P = E / T$, where P is the constant power, needed to convert energy E *(perform work)* during period of duration T. Energy is measured in joules (J). Power is meassured in watts (W) which is equal to one joule per second (J/s). It is important to stress that energy is also subject to the *law of conservation of energy*. According to this law, energy can neither be created nor destroyed by itself. It can only be transformed. Hence, according to the energy conservation law, the total inflow of energy into a system must equal the total outflow of energy from the system, plus the change in the energy contained within the system. In our setting, this law means that a mobile device i cannot use more power (energy over time) than power available through the device battery which stores energy as an *energy reserve* E_i. We will apply this law as we formulate the power consumption constraints for the traffic routing problem.

In the traffic routing problem, we want to know and control *power consumption* on a device in terms of how much energy (work) a device (e.g., phone) uses, when receiving and transmitting traffic over the wireless network. The power consumption very much depends on the *radio frequency propagation model.* The radio propagation model is an empirical mathematical formulation for the characterization of radio wave propagation. It is expressed as a *function of frequency, distance, and other wireless channel conditions.* A *single model* is usually developed to predict the behavior of propagation for all similar links under similar constraints which is our case. We will use a single radio propagation model for all links in our wireless networks. Furthermore, we will use a simple first order function for representing the radio propagation where this function will depend on *local device* and on the *distance* how far we are transmitting from one device to another. Survey of radio propagation models can be found in [120, 121].

The power consumption, applied to the simple radio propagation model, will then give us the *power usage* when *transmitting* and *receiving bits* from/to source to/from destination, and this power consumption will be measured in units *joules per bit (J/bit)*. Let p_i^r (J/bit) be the power consumption at node i, when it receives one unit of data, and p_{ik}^t(J/bit) be the power consumption when one unit of data is transmitted from i over link (i,k). Then, we will express the power consumption p_i^r at the *receiving* node $\boldsymbol{i}$ as *a constant usage of power* at the device when receiving a packet. The power consumption p_{ik}^t at the *sending* node $\boldsymbol{i}$ that sends packet to a destination k consists of (a) *sender's device constant usage of power* to process locally the sending packet, and (b) *sender's device distance* to the destination $\boldsymbol{k}$ which is propotional to the amplifier energy needed to send the packet from node i to node k. Equations in Equation 5.4 express the power consumption for sending and receiving a packet, related to the single radio propagation model

$$p_i^r = \alpha \text{ and } p_{ik}^t = \alpha + \beta \cdot d_{ik}^m , \tag{5.4}$$

where α is a *distance-independent* constant that represents the *energy consumption* to run the *transmitter or receiver circuitry.* β is the coefficient of the *distance-dependent* term that represents the transmit amplifier energy. The exponent m is determined from field measurements and it is typically a constant between 2 and 4. Note that Equation 5.4 is an assumption in the presented routing framework, i.e., what kind of radio propagation model we are using. Other, more complex and more accurate radio propagation models (e.g., models using frequencies, distance, interference) can be used which might provide more accurate specification for power consumption when transmitting and receiving bits.

Now that we have described how much power will be *consumed per bits*, when transmitting and receiving in the wireless network, we derive the traffic power consumption at a node $\boldsymbol{i}$ per time interval (in J/s).

The *power consumption at node i* will consist of two major terms:

a. power consumption of *outgoing transmitted traffic*, i.e., the *routing variables* $\phi_{ik}(j)$ for all destinations j (representing the *outgoing transmitting fraction of traffic*) and the *total node flow rate* $t_i(j)$ for all destinations j jointly determine the traffic ***sent*** from node $\boldsymbol{i}$ along wireless link (i,k);
b. power consumption of *incoming receiving traffic*, i.e., the *total node flow rate* $t_i(j)$ for all destinations j, and the *incoming receiving traffic rate* $r_i(j)$ for all destinations j, jointly determine the traffic ***received*** at node $\boldsymbol{i}$.

We can then express the power consumption p_i (in J/s) at the wireless node i as follows:

$$p_i = \sum_{j \in N} [t_i(j) \sum_{k \in N} p_{ik}^t \phi_{ik}(j) + p_i^r (t_i(j) - r_i(j))] \tag{5.5}$$

Where the term $t_i(j) \sum_{k \in N} p_{ik}^t \phi_{ik}(j)$ represents the power consumption at node $\boldsymbol{i}$ for *transmitting outgoing traffic* to all neighboring nodes k that are along the route to destination j; and the term $p_i^r (t_i(j) - r_i(j))$ represents the power consumption at node $\boldsymbol{i}$ for *receiving incoming traffic* $t_i(j)$ going to destination $\boldsymbol{j}$ minus the local traffic $r_i(j)$ that is generated at the node $\boldsymbol{i}$ (for local generated traffic we do not spend any receiving energy, only sending energy).

5.4.3 Maximum Network Lifetime Routing Problem

Under the assumptions of the *network, traffic demand* and power consumption descriptions, described in Sections 5.4.1 and 5.4.2, we will now consider the *routing design* with the "system optimization" of the overall network in mind in which the entire network properly functions. It means, we want to find optimal routes so that the overall lifetime of the network is maximized under energy conservation, flow conservation, and power consumption constraints. Note that the network lifetime T in the presented design is understood as the lifetime T_i of the wireless node which dies first, i.e., $T = \min_{i \in N} T_i$.

To find the best routes, we need to solve an *optimization problem* which can be formulated as follows. The routing optimization problem asks that given network and traffic demand, how to route the traffic so that the network lifetime T (QoS parameter) can be *maximized.* If we assume T_i to be the lifetime of wireless node i, and T the *lifetime of the wireless network*, then the routing problem can be formulated as a *linear optimization problem* **T** with three constraints: (a) energy conservation, (b) flow conservation and (c) power consumption constraints:

T: maximize T (maximize lifetime of given wireless network)
subject to

$$p_i \cdot T \leq E_i, \forall i \in N \quad \text{(energy conservation constraint)} \tag{5.6}$$

$$t_i(j) = r_i(j) + \sum_{l \in N} t_l(j)\phi_{li}(j), \forall i, j \in N \quad \text{(flow conservation constraint)}$$

$$p_i = \sum_{j \in N} [t_i(j) \sum_{k \in N} p_{ik}^t \phi_{ik}(j) + p_i^r (t_i(j) - r_i(j))] \quad \text{(power consumption constraint)}$$

Equation 5.6 represents *the energy conservation constraint*, and it comes from the fact that the energy consumption at any node i within the network lifetime is no more than its *energy reserve* E_i, i.e., $p_i \cdot T \leq p_i \cdot T_i = E_i, \forall i \in N$. It means we cannot use up more power p_i over time T than what energy reserve E_i we have at node i. Power usage p_i at node i is constrained by the power consumption constraint of the *incoming* and *outgoing traffic* at the node i as shown in Equation 5.5, and the traffic at node i is constrained by the *flow conservation* as shown in Equation 5.3.

The result of solving the problem **T** for a given network (N,L) and *end-to-end traffic rates* $r_i(j)$ is the *optimal routing solution* $\Phi = \{\phi_{ik}(j) \mid i, j, k, \in N\}$. Φ represents the routing set of routes and their fractions of traffic going along the routes. We can acquire the optimal routing solution Φ by solving the above *linear optimization problem* via a *centralized* algorithm [e.g., 62]). However, the real challenge is how to solve this routing problem **T** in a *distributed fashion*.

To address this challenge, we will transform the routing optimization problem **T** into a *utility-based* nonlinear optimization formulation of the maximum lifetime routing problem. This transformation will then lead to a fully distributed routing algorithm. The *utility-based routing problem formulation* is inspired by the *max–min resource allocation problem* in distributed computing and networking area which we have also discussed in detail in Chapter 3.3.1. Recall that *max–min fairness* means that for any user i, increasing its resource share x_i cannot be achieved without decreasing the resource share of another user x_j that satisfies $x_i \geq x_j$. Simply put, max–min allocation mechanism *maximizes the resource share* of the user who is allocated with the *minimum resource.*

In the context of the maximum lifetime routing problem, if we regard *lifetime* T_i of a node as a *"resource"* of its own, then the goal of maximizing network lifetime is actually to "*allocate lifetime*" to each node so that *max–min fairness* is satisfied. This "lifetime allocation" mechanism needs to be achieved via *routing* and has to satisfy the energy constraint, flow conservation constraint and power consumption constraint.

We further adopt the concept of "utility" which is being widely used in the area of resource management. We have discussed extensively utilities and utility-based optimization problems in resource management in Chapters 2 and 3. Recall that each resource has its own *utility*. Defined on the resource share of a user, utility usually represents the degree of satisfaction of this user. By defining an appropriate *utility function*, the problem of achieving max-min fairness can be converted into the problem of *maximizing the aggregated utility* (sum of utilities of all users) [14, 63].

If we define utility U_i of a node i as a function of its "*resource*" lifetime T_i, then the routing problem **T** can be transformed and reformulated to the routing problem **U** of maximizing the aggregated utility of all nodes within the network which can be solved in a distributed fashion.

U: $\quad$ *maximize* $U = \sum_{i \in \mathrm{N}} U_i$

subject to

$p_i \cdot T \leq E_i, \forall i \in \mathrm{N}$ (energy conservation constraint)

$t_i(j) = r_i(j) + \sum_{l \in \mathrm{N}} t_l(j)\phi_{li}(j), \forall i, j \in \mathrm{N}$ (flow conservation constraint)

$p_i = \sum_{j \in \mathrm{N}} [t_i(j) \sum_{k \in \mathrm{N}} p^t_{ik}\phi_{ik}(j) + p^r_i(t_i(j) - r_i(j))]$ (power consumption constraint)

To solve the routing problem **U**, several assumptions need to be made.

First, U_i of a node i is defined as a *function of its lifetime* as follows:

$$U_i(T_i) = \frac{T_i^{1-\gamma}}{1-\gamma} . \gamma \rightarrow \infty \tag{5.7}$$

where $T_i = E_i / p_i$ and γ can be made arbitrarily large to infinite. How to determine the value of γ and its impact will be briefly discussed in Section 5.4.5 as one of the practical considerations of the routing approach. Note that it was shown in Referenes [14, 63] that $T_i, i \in N$ satisfies *max–min fairness* if and only if it solves the aggregated utility maximization problem $\max \sum_{i \in N} U_i$ with U_i defined as in Equation 5.7.

Second, *optimality conditions* (e.g., maximum of the aggregate utility $U = \sum_{i \in N} U_i$ exists, partial derivatives of U, $\partial U / \partial r_i(j)$ exist) in problem **U** must be satisfied. The discussion about optimality conditions of the routing problem **U** can be found in [42].

5.4.4 Distributed Routing

With the presented *utility-based problem formulation*, we now derive a *distributed routing* algorithm that solves the routing problem **U** in order to achieve maximum lifetime routing.

The distributed algorithm works in an *iterative* fashion. In each iteration, for each node i and a given *commodity* j (flow routed to destination j), node i must incrementally decrease the *fraction of traffic* on link (*i,k*) (by decreasing $\phi_{ik}(j)$), whose *marginal utility* $\delta_{ik}(j)$ is large, and do the reverse for those links whose marginal utility is small, until the marginal utilities of all links carrying traffic are equal. When this condition is met for all nodes and all commodities (flows), the entire system reaches the optimal point. Note that *marginal utility* of a service is a utility (degree of satisfaction) gained (or lost) from an increase (or decrease) in the consumption of that service.

Therefore, for each node i, each iteration involves two steps: (1) the *calculation of the marginal utility* for each outgoing link *(i,k)* and each of its downstream neighboars *k*'s marginal utility; and (2) the *adjustment of routing variables* $\phi_{ik}(j)$ based on marginal utilities.

Calculation of marginal utilities. To derive the marginal utilities we make first several observations:

First, since the utility function U_i is a function of the node lifetime T_i, (due to Equation 5.7), which directly associates with p_i (due to $T_i = E_i / p_i$), we can rewrite the utility function $U_i(T_i)$ as $U_i(p_i)$ with U_i being a function of p_i.

Second, power consumption p_i depends on the end-to-end traffic set $\mathbf{r} = \{r_i(j) \mid i, j \in N\}$ and the routing set $\mathbf{\Phi} = \{\phi_{ik}(j) \mid i, j, k \in N\}$, hence the aggregate utility U depends on $\mathbf{r}$ and $\mathbf{\Phi}$. This means that to solve the optimization routing problem $\mathbf{U}$, we then need to consider the calculation of partial derivates of the aggregate utility U with respect to r and the routing variables ϕ, respectively.

Third, if we assume that U_i is *concave and continuously differentiable* for p_i, $\forall i$, then U is *maximized* if and only if for $\forall i, j \in$ N:

$$\frac{\partial U}{\partial r_i(j)} \begin{cases} = \delta_{ik}(j), if\ \phi_{ik} > 0 \\ \geq \delta_{ik}(j), if\ \phi_{ik} = 0 \end{cases} \tag{5.8}$$

where $\frac{\partial U}{\partial r_i(j)}$ is called the *marginal utility of node i*, $\delta_{ik}(j) = U'_{ik} + \frac{\partial U}{\partial r_k(j)}$ is called the *marginal utility on link* (i,k) *with respect to commodity* $\boldsymbol{j}$, and U'_{ik} is called the *marginal utility on link* (i,k). This statement (Equation 5.8) states that the aggregate utility is maximized if at any node i, for a given commodity j (flow destined to j),

1. all links (i,k) that have any portion of flow rate $t_i(j)$ routed through $(\phi_{ik}(j) > 0)$ must achieve the same marginal utility δ_{ik} with respect to j, and
2. this maximum marginal utility must be greater than or equal to the marginal utilities δ_{ik} of the links with no flow routed $(\phi_{ik}(j) = 0)$.

We first introduce the calculation of the *link marginal utility* U'_{ik}. The link marginal utility is $U'_{ik} = p^t_{ik} U'_i(p_i) + p^r_k U'_k(p_k)$ because sending data and receiving data over a wireless link *(i,k)* requires power consumption of both sending node i and receiving node k. Thus, the calculation of U'_{ik} depends on cooperation of both nodes. The protocol for calculating U'_{ik} is as follows:

(a) Node i is responsible to calculate the term $p^t_{ik} U'_i(p_i)$. $U'_i(p_i)$ can be derived based on (Equation 5.7) if the energy reserve E_i and power consumption p_i are known. Both values can be directly measured by node i. p^t_{ik} can be calculated based on (Equation 5.4), if constants α, β, m, and node distance d_{ik} are known beforehand. Alternatively, node i can directly estimate p^t_{ik} by measuring the amount of data sent from i to k and corresponding power consumption.

(b) Node k is responsible to calculate the term $p^r_k U'_k(p_k)$. $U'_k(p_k)$ can be calculated the same way as $U'_i(p_i)$. p^r_k can be either calculated based on Equation 5.4 or directly estimated by measuring the amount of data received at node k and the corresponding power consumption. After calculation, j can send the value of $p_k^r U_k'(p_k)$ to node i, which in turn acquires U_{ik}'.

Next, at node i we calculate the marginal utility of link (i,k) with respect to commodity j, as follows: $\delta_{ik}(j) = U_{ik}' + \dfrac{\partial U}{\partial r_k(j)}$ is the marginal utility of all its outgoing links regarding commodity j. We have discussed how to calculate U_{ik}', and $\partial U / \partial r_k(j)$ is the marginal utility of i's downstream neighbor k.

As last, we calculate node i's *marginal utility* $\partial U / \partial r_i(j)$ according to Equation 5.8. This marginal utility is calculated in a *recursive way*. Starting from node j, the recipient of commodity j will have $\partial U / \partial r_j(j) = 0$, based on the definition of Φ (since $\phi_{jk}(j) = 0$). j then sends the value of $\partial U / \partial r_j(j)$ and $p_j^r U_j'(p_j)$ to its upstream neighbor, say k. Upon receiving the updates, node k can calculate U_{ik}' as described above, then acquire $\partial U / \partial r_k(j)$. Then, k repeats the same procedure to its upstream neighbor, until node i is reached.

In summary, for each node i, each routing iteration (k) involves the following steps:

1. Node i calculates the marginal utility U_{ik}' for each outgoing link (i,k), gets updates of marginal utility $\partial U / \partial r_k(j)$ from each of its downstream neighbors k, and then calculates $\delta_{ik}(k)$, the marginal utility of link (i,k) with respect to commodity j.
2. Node i calculates its own marginal utility $\partial U / \partial r_i(j)$ as discussed above and sends it to its upstream neighbors.
3. Node i adjusts *routing variables* $\phi_{ik}(j)$, based on the values of marginal utilities for each outgoing link (i,k) and for downstream neighbors k with respect to the commodity j. Variable $\phi_{ik}(j)$ value then determines the routing and traffic fractions to the next node. The adjustment happens as follows:

    a. Calculate vector of routing changes $\Delta\phi_i^{(k)}(j) = (\Delta\phi_{i1}^{(k)}(j), \ldots, \Delta\phi_{im}^{(k)})$, which represents changes made to the routing solution $\phi^{(k)}(j)$ that we got during the iteration (k):

$$\begin{cases} \Delta\phi_{il}^{(k)}(j) = -\min\{\phi_{il}^{(k)}(j), \dfrac{\rho(\delta_{il}(j) - \delta_{\min}(j))}{t_i(j)}\}; & \delta_{il}(j) \neq \delta_{\min}(j) \\ \sum_{\delta_{im}(j) \neq \delta_{\min}(j)} \Delta\phi_{im}^{(k)}(j); & \delta_{il}(j) = \delta_{\min}(j) \end{cases} \tag{5.9}$$

where $\delta_{\min}(j) = \min_{m \notin B_{i,\phi^{(k)}(j)}} \delta_{im}(j)$, $\rho > 0$ is some positive step size. $B_{i,\phi}(j)$ is the *set of blocked nodes* k for which $\phi_{ik}(j) = 0$, and the algorithm is not permitted to increase $\phi_{ik}(j)$ from 0. $k \in B_{i,\phi}(j)$ if one of the following conditions is met:

(1) k is not a neighbor of i, i.e., $(i,k) \notin L$;

(2) the marginal utility of k is already greater than or equal to the marginal utility of i, i.e., $\phi_{ik}(j) = 0$ and $\partial U/\partial r_i(j) \geq \partial U/\partial r_k(j)$.

(3) link (l,m) is an improper link, i.e., $\phi_{ik}(j) = 0$ and $\exists(l,m) \in L$ such that (a) $l = k$ or l is downstream to k with respect to commodity j; (b) $\phi_{lm}(j) > 0$, and $\partial U/\partial r_l(j) \geq \partial U/\partial r_m(j)$.

Note that the set $B_{i\phi}(j)$ ensures loop-free routing, for each node i, with respect to commodity j.

b. Adjust routing variables $\phi_i^{(k+1)}(j) = \phi_i^{(k)}(j) + \Delta\ \phi_i^{(k)}(j), \forall i \in N - \{j\}$.

5.4.5 Practical Issues

Static wireless ad hoc networks can encompass two types of networking scenarios: sensor networks and ad hoc networks. In the scenario of sensor networks, usually one node is picked as the base station (data sink) and a subset of the other nodes acts as data sources sending traffic to the sink. The rest of the nodes act as relaying nodes. In the ad hoc network scenario, there exist random pairs of unicast connections between senders and receivers. The rest of nodes serve as *relay nodes*. We present couple of observations of energy-efficient routing in these two scenarios:

First, *maximum lifetime of the network* drops at a super linear speed as the number of data sources increases as shown in Figure 5.13. The decrease is mainly as a result of increased traffic demand in a network with fixed overall *energy reserve*. The figure shows comparison of four algorithms via simulation with maximal transmission range 25 m of each node, $\alpha = 50$ nJ/b, $\beta = 0.0013$ pJ/b/m^4, m = 4 for power consumption model, $E_i = 50$ kJ, sending rate of each connection 0.5 Kbps: (a) *optimal MaxLife* result represents the centralized solution of the problem **T**; (b) *MaxLife* (with $\gamma = 3$ and $\gamma = 4$) represents the distributed solution of the problem **U** with different utility values $U_i(T_i)$ influenced by the internal parameter γ; (c) *MinEnergy* result represents an algorithm that minimizes the energy consumption for each data unit routed through the network. For each data source, the algorithm finds its shortest path to the destination in terms of energy cost. The route for each data source is fixed throughout the entire network lifetime.

Second, as Figure 5.13 shows, in the sensor network scenario, the *MinEnergy* algorithm has the worst performance. The reason is that *MinEnergy* ends up with a shortest path data aggregation tree from data sources to data sink, where the entire traffic concentrates on a few nodes located close to the data sink. The energy reserve of these nodes can easily run out very soon, which is the main reason for the inferior performance of this algorithm. The same traffic concentration problem exists in ad hoc network scenario, but not as serious as the previous scenario. In contrast, the MaxLife

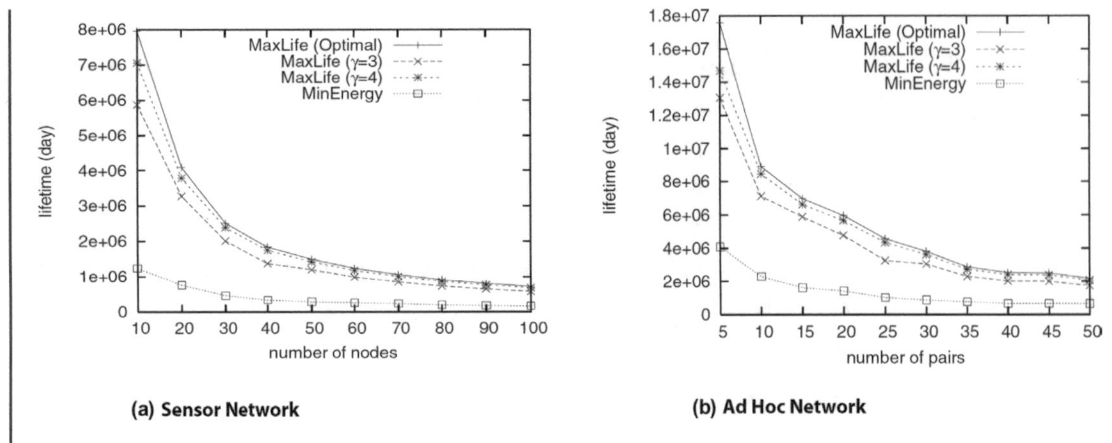

FIGURE 5.13: Network lifetime of networks in two scenarios [42].

algorithms are able to effectively diverge the traffic, hence keep energy consumption lower among all nodes, which significantly prolongs the network lifetime.

Third, the *MaxLife* algorithm consumes more *system energy* than MinEnergy algorithm for an average bit of data routed through the network as shown in Figure 5.14. The reason is that in order to *maximally utilize the energy reserve* of all nodes within the network, sometimes the data from a source has to go through some route whose energy consumption rate is not as efficient as the one returned by *MinEnergy* algorithm.

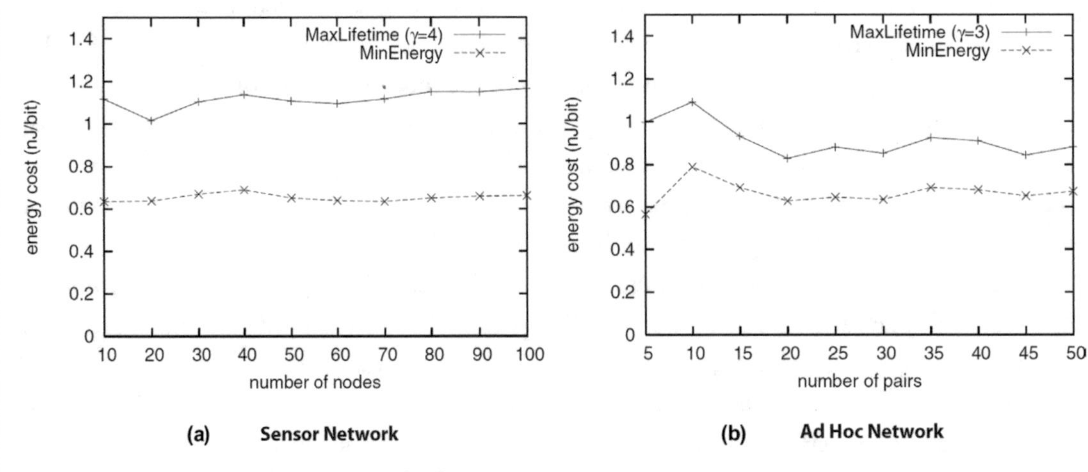

FIGURE 5.14: Average energy cost [42].

5.5 SUMMARY

In this chapter, we have shown three routing approaches that are relevant to the current static and mobile wireless environment.

The first algorithm in Section 5.2 aims to make readers aware that routing, and especially QoS routing very much benefits from *cross-layer design* and from *additional contextual information* such as location. Since location acquisition is becoming pervasive on current phones (outdoor and indoor), and various sensors on mobile phones enable us to identify group mobility patterns and directions of movement [77, 78, 79], moving toward *predictive location-based routing* approach for group-based data accessibility is feasible for group-based scenarios.

In Section 5.3, we have shown the design of a routing algorithm and protocol in harsher environments that can be present due to *overloaded* or *adversarial situations*. This approach concentrates on *fault-tolerant routing algorithm and protocol design*. We show that if one wants to select fault tolerant routes with QoS requirements, one needs to (a) have status information of the network/nodes available, and (b) carefully decide which route to select according to wireless situational awareness and QoS requirements.

When considering routing of data with QoS performance in mind, one of the metrics is *energy*, an important resource for mobile nodes such as sensors, phones, tablets or laptops. In Section 5.4, we discussed a more theoretical approach to designing a distributed energy-constrained routing algorithm. We wanted to show how to formulate an energy routing problem as a "system optimization" problem to maximize the *network lifetime* under energy conservation, flow conservation and power consumption constraints in static multi-hop wireless networks with QoS requirements.

As we mentioned in the introduction, the routing space is very large due to the diversity of the environments in which wireless and mobile nodes exist, and the readers are recommended to carefully pick routing approaches depending on the *physical environments, mobile node capabilities* in terms of its battery lifetime, wireless network capacilities and computational power, and *application QoS requirements* that are coming from the distributed applications as well from the physical constraints of the tasks.

• • • •

Acknowledgment

This lecture summarizes work of many researchers, but especially the joint research work and lessons learned that I did with my former and current PhD students Baochun Li, Kai Chen, Wenbo He, Hoang Ngyuen, Raoul Rivas, Samarth Shah, and Yuan Xue. I would like to acknowledge and thank them for their research collaboration and many contributions to the difficult task of exploring QoS in wireless networks over unlicensed spectrum. I would like to acknowledge and thank my editor, Professor M. Satyanarayan, who provided very insightful feedback to each chapter of the lecture and worked tirelessly with me to bring the lecture to a much better and understandable level for a broader audience. I would like to acknowledge and thank Michael Morgan for his assistance toward the completion of the overall book. My current PhD students Hongyan Li and Yunlong Gao assisted me in proofreading the lecture and providing me with readibility feedback on the final manuscript. Last, but not the least, I would like to thank my husband Normand for his incredible patience, encouragement, and love during the whole writing process and other members of my family, Ruzena, Peter Jr., Jill, Peter Sr., Zuzana, Andrea, and Martin who were always there for me with their loving support as I moved from one stage to another in the writing process. Thank you all!

References

[1] IEEE Standard 802.11, "Wireless LAN medium access control (MAC) and physical layer (PHY) specification," 1999.

[2] Xue, Y., Chen K., and Nahrstedt, K., "Achieving proportional delay differentiation in wireless LAN via cross-layer scheduling," *Wireless Communications and Mobile Computing Journal* 4: 849–866, Wisley InterScience Publisher, 2004. doi:10.1002/wcm.259

[3] Yang, Y., and Kravets, R., "Distributed QoS Guarantees for Real-time Traffic in Ad Hoc Networks," IEEE 1st Annual IEEE Communications Society Conference on Sensor and Ad Hoc Communications and Networks (SECON), 2004.

[4] Selten, F.M., "Baro-clinic empirical orthogonal functions as basic functions in an atmospheric model," *J. Atmos. Sci.* 54, 2099–2114, 1997. doi:10.1175/1520-0469(1997)054<2099:BEOFAB>2.0.CO;2

[5] Timmermann, A., Voss, H.U., and Pasmanter, R. 2001. "Empirical dynamical system modeling of ENSO using non-linear inverse techniques." *J. Phys. Oceanogr.* 31, 1579–1598. doi:10.1175/1520-0485(2001)031<1579:EDSMOE>2.0.CO;2

[6] Hou, I.-H., "Delay-Constrained Throughput In Wireless Networks: Providing Reliable Services over Unreliable Channels," Thesis Proposal, 2009, University of Illinois at Urbana-Champaign.

[7] Dovrolis, C., Ramanathan, P., "A case for relative differentiated services and the proportional differentiation model," IEEE Network, 1999, 13(5): 26–34. doi:10.1109/65.793688

[8] He, W., Nahrstedt, K., "Impact of Upper Layer Adaptation on End-to-End Delay Management in Wireless Ad Hoc Networks," 12th IEEE Real-Time and Embedded Technology and Applications Symposium, April 2006.

[9] Shah, S., Chen, K., Nahrstedt, K., "Dynamic Bandwidth Management in Single-Hop Ad Hoc Wireless Networks," Mobile Networks and Applications Journal, Kluwer Academic Publishers, 10, 199–217, 2005. doi:10.1023/B:MONE.0000048555.72514.9a

[10] Gupta, P. and Kumar, P.R., "The capacity of wireless networks," IEEE Transactions on Information Theory, Vol. 46, No. 2, pp. 388–404, March 2000. doi:10.1109/18.825799

[11] Luo, H., and Bhargavan, V. "A New Model for Packet Scheduling in Multihop Wireless Networks," ACM Mobihoc, 2000, pp. 76–86. doi:10.1145/345910.345923

[12] Tassiulas, L. and Sarkar, S., "Max-min fair scheduling in wireless networks," IEEE INFOCOM 2002, pp. 763–772. doi:10.1109/INFCOM.2002.1019322

[13] Xue, Y., Li, B., Nahrstedt, K., "Optimal Resource Allocation in Wireless Ad Hoc Networks: A Price-based Approach," IEEE Transactions on Mobile Computing, Vol. 5, Issue 4, April 2006.

[14] Kelly, F.P., "Charging and Rate Control for Elastic Traffic." European Transactions on Telecommunications, vol. 8, pp. 33–37, 1997. doi:10.1002/ett.4460080106

[15] Kelly, F.P., Maulloo, A.K., Tan, D.K.H., "Rate Control in Communication Networks: Shadow prices, Proportional Fairness and Stability," Journal of the Operational Research Society, vol. 49, pp. 237–252, 1998. doi:10.1038/sj.jors.2600523; doi:10.2307/3010473; doi:10.1057/palgrave.jors.2600523

[16] Augustson, J.G., Minker, J., "An Analysis of Some Graph Theoretical Cluster Techniques," Journal of Association for Computing Machinery, vol. 17, no. 4, pp. 571–586, 1970. doi:10.1145/321607.321608

[17] Bertsekas, D., Gallager, R., "Data Networks," 2nd edition, Chapter 6, Prentice Hall, 1996.

[18] Ahn, G., Campbell, A., Veres, A., Sun, L., "Service differentiation in stateless wireless ad hoc networks," IEEE INFOCOM 2002, New York, NY.

[19] Liao, R., Wouhaybi, R., Campbell, A., "Incentive engineering in wireless LAN-based access networks," IEEE ICNP 2002, Paris, France. doi:10.1109/ICNP.2002.1181386

[20] Marbach, P., Berry, R., "Downlink resource allocation and pricing for wireless networks," IEEE INFOCOM 2002, New York, NY. doi:10.1109/INFCOM.2002.1019398

[21] Qiu, Y., Marbach, P., "Bandwidth allocation in ad-hoc networks: a price-based approach," IEEE INFOCOM 2003, San Francisco, CA. doi:10.1109/INFCOM.2003.1208917

[22] Grossglauser, M., Keshav, S., Tse, D., "Rcbr: A simple and efficient service for multiple time-scale traffic," ACM SIGCOMM 1995, Cambridge, MA. doi:10.1145/217391.217436

[23] Ramanathan, P., Dovrolis, C., Stiliadis, D., "Proportional differentiated services: Delay differentiation and packet scheduling," IEEE/ACM Transactions on Networking, vol. 10, pp. 12–26, Feb. 2002. doi:10.1109/90.986503

[24] Almquist, A., Lindgren, A., Schelen, O., "Quality of Service schemes for IEEE 802.11 wireless LANs—an evaluation," Mobile Networks and Applications, vol. 8., no. 3, pp. 223–235, June 2003. doi:10.1023/A:1023389530496

[25] Kanodia, V., Li, C., Sabharwal, A., Sadeghi, B., Knightly, E., "Distributed priority scheduling and medium access in ad hoc networks," Wireless Networks, vol. 8., no. 1, November 2002. doi:10.1023/A:1016538128311

[26] Yang, X., Vaidya, N.H., "Priority scheduling in wireless ad hoc networks," ACM Mobicom, 2002. doi:10.1145/513800.513809

[27] Aad, I., Castelluccia, C., "Differentiation mechanisms for IEEE 802.11," IEEE INFOCOM 2001. doi:10.1109/INFCOM.2001.916703

[28] Xue, Y., Chen, K., Nahrstedt, K., "Distributed End-to-End Proportional Delay Differentiation in Wireless LAN," IEEE ICC, 2004.

[29] He, W., Nahrstedt, K., "Impact of Upper Layer Adaptation on End-to-end Delay Management in Wireless Ad Hoc Networks," IEEE RTAS 2006.

[30] Ngyuen, H., Rivas, R., Nahrstedt, K., "iDSRT: Integrated Dynamic Soft Real-time Architecture for Critical Infrastructure Data Delivery over WLAN," ICST QShine 2009. doi:10.1007/978-3-642-10625-5_12

[31] Ljung, L., "System Identification: Theory for the User," 2nd edition, Prentice Hall, 1999.

[32] Liu, C.L., Layland, J.W., "Scheduling algorithms for multiprogramming in a hard-real-time environment," *Journal of ACM*, 20(1): 46–61, 1973. doi:10.1145/321738.321743

[33] Crenshaw, T.L., Hoke, S., Tirumala, A., Caccamo, M., "Robust implicit EDF: a wireless MAC protocol for collaborative real-time systems," IEEE Transactions on Embedded Computing Systems, 2007. doi:10.1145/1274858.1274866

[34] Caccamo, M., Zhang, L.Y., Sha, L., Buttazzo, G., "An implicit prioritized access protocol for wireless sensor networks," IRRR Real-time Systems Symposium (RTSS), 2002. doi:10.1109/REAL.2002.1181560

[35] Gopalan, K., Kang, K.-D., "Coordinated allocation and scheduling of multiple resources in real-time operating systems," Workshop on Operating Systems Platforms for Embedded Real-Time Operating Systems, 2007.

[36] Gopalan, K., Cker Chiueh, T., "Multi-resource allocation and scheduling for periodic soft-real-time applications," ACM/SPIE Multimedia Computing and Networking, 2002.

[37] Ghosh, S., Rajkumar, R., Lehoczky, J., "Integrated resource management and scheduling with multi-resource constraints," IEEE Real-Time Systems Symposium (RTSS), 2004. doi:10.1109/REAL.2004.25

[38] Xu, D., Nahrstedt, K., Viswanathan, A., Wichadakul, D., "QoS and Contention-aware multi-resource reservation," IEEE International Symposium on High Performance Distributed Computing (HPDC), 2000. doi:10.1109/HPDC.2000.868629

[39] Buttazzo, G., "Hard Real-Time Computing Systems, Predictable Scheduling Algorithms and Applications," Kluwer Academic Publishers, 1997.

[40] Baruah, S., Rosier, L., Howell, R.R., "Algorithms and complexity concerning the preemptive scheduling of periodic, real-time tasks on one processor," Journal of Real-Time Systems, 1990. doi:10.1007/BF01995675

[41] Bertsekas, D.P., "Non-linear Programming," Athena Scientific Publishers, 1999.

[42] Xue, Y., Cui, Y., Nahrstedt, K., "A Utility-based Distributed Maximum Lifetime Routing

Algorithm for Wireless Networks," IEEE Transactions on Vehicular Technology, vol. 55, no. 3, pp. 797–805, 2006.

[43] Stojmenovic, I., Lin, X., "Power-aware Localized Routing in Wireless Networks," IEEE Transactions on Parallel and Distributed Systems, 12(11): 1122–1133, 2001. doi:10.1109/71.969123

[44] Singh, S., Woo, M., Raghavendra, C., "Power-aware Routing in Mobile Ad hoc Networks," ACM MOBICOM 1998. doi:10.1145/288235.288286

[45] Rodoplu, V., Meng, T., "Minimum Energy Mobile Wireless Networks," IEEE Journal of Selected Areas in Communications, 17(8): 1333–1344, 1999. doi:10.1109/49.779917

[46] Gomez, J., Campbell, A., Naghshinen, M., Bisdikian, C., "Conserving Transmission Power in Wireless Ad Hoc Networks," 9th International Conference on Network Protocols (ICNP), 2001. doi:10.1109/ICNP.2001.992757

[47] Xue, Y., Nahrstedt, K., "Providing Fault-Tolerant Ad Hoc Routing Service in Adversarial Environments," Kluwer Academic Publishers, *Wireless Personal Communications Journal*, 29: 367–388, 2004. doi:10.1023/B:WIRE.0000047071.75971.cd

[48] Johnson, D.B., Maltz, D.A., "Dynamic Source Routing in Ad Wireless Networks," Mobile Computing, pp. 153–181, 1996. doi:10.1007/978-0-585-29603-6_5

[49] Yi, S., Nalburg, P., Kravets, R., "Security-aware Ad hoc routing for wireless networks," Technical Report, No: UIUCDCS-R-2001-2241, UILU-ENG-2001-1748, 2001.

[50] Papadimitratos, P., Haas, Z.J., "Secure Routing for Mobile Ad Hoc Networks," SCS Communication Networks and Distributed Systems Modeling and Simulation Conference, 2002.

[51] Hu, Y.-C., Perrig, A., Johnson, D.B., "Ariadne: A Secure on-Demand Routing Protocol for Ad Hoc Networks," ACM MOBICOM 2002.

[52] Castaneda, R., Das, S.R., "Query Localization Techniques for On-Demand Routing Protocols in Ad Hoc Networks," ACM MOBICOM 1999.

[53] Ko, Y., Vaidya, N., "Location-aided Routing (LAR) in Mobile Ad Hoc Networks," ACM MOBICOM 1998. doi:10.1145/288235.288252

[54] Chen, K., Shah, S., Nahrstedt, K., "Cross-Layer Design for Data Accessibility in Mobile Ad Hoc Networks," Kluwer Academic Publishers, Wireless Personal Communications 21: 49–76, 2002. doi:10.1023/A:1015509521662

[55] Hu, Y.C., Johnson, D.B., "Catching Strategies in On-Demand Routing Protocols for Wireless Ad Hoc Networks," ACM MOBICOM 2000.

[56] Liang, B., Haas, Z.J., "Predictive Distance-based Mobility Management for PCS Networks," IEEE INFOCOM 1999. doi:10.1109/INFCOM.1999.752157

[57] Chen, S., Nahrstedt, K., "Distributed Quality of Service Routing in Ad Hoc Networks," IEEE Journal of Selected Areas in Communication, 17(8), 1999.

[58] Lee, S.B., Ahn, G.S., Zhang, X., Campbell, A., INSIGNIA: An IP-based Quality of Service Framework for Mobile Ad Hoc Networks," Journal of Parallel and Distributed Computing, Vol. 60, pp. 374–406, 2000. doi:10.1006/jpdc.1999.1613

[59] Chen, K., Nahrstedt, K., "Effective Location-Guided Overlay Multicast in Mobile Ad Hoc Networks," International Journal of Wireless and Mobile Computing, Special Issue on Group Communications, Vol. 3, 2005.

[60] Chen, K., Nahrstedt, K., "Effective location-guided tree construction for small group multicast in MANET," IEEE INFOCOM 2002.

[61] Lindgren, A., Doria, A., "Probabilistic Routing Protocol for Intermittently Connected Networks," IETF draft-irtf-dtnrg-prophet-01, November 17, 2008.

[62] Garg, N., Konemann, J., "Faster and Simpler Algorithms for Multi-commodity Flow and Other Fractional Packing Problems," IEEE Symposium on Foundations in Computer Science, pp. 300–309, 1998. doi:10.1109/SFCS.1998.743463

[63] Srikant, R., "The Mathematics of Internet Congestion Control," Birkhauser, 2004.

[64] Zhang, Y., Lee, W., "Intrusion Detection in Wireless Ad Hoc Networks," ACM MOBICOM 2000. doi:10.1145/345910.345958

[65] Buchegger, S., Boudec, J.Y.L., "Performance Analysis of the CONFIDANT Protocol," ACM MobiHoc, 2002.

[66] Michiardi, P., Molva, R., "Core: A Collaborative Reputation Mechanism to Enforce Node Cooperation in Mobile Ad Hoc Networks," Proceedings of Communications and Multimedia Security, 2002.

[67] Buttyan, L., Hubaux, J.P., "Enforcing Service Availability in Mobile Ad Hoc WANs," IEEE/ACM MobiHoc, 2002. doi:10.1109/MOBHOC.2000.869216

[68] Buttyan, L., Hubaux, J.P., "Simulating Cooperation in Self-Organizing Mobile Ad Hoc Networks" ACM/Kluwer Mobile Networks and Applications, Vol. 8, no. 5, 2003.

[69] Sundaramurthy, S., Belding-Royer, E.M., "The AD-MIX Protocol for Encouraging Participation in Mobile Ad Hoc Networks," IEEE ICNP, 2003. doi:10.1109/ICNP.2003.1249765

[70] Kong, J., Zerfos, P., Luo, H., Lu, S., Zhang, L., "Providing Robust and Ubiquitous Security Support for Mobile Ad Hoc Networks," IEEE ICNP, 2001.

[71] Zhou, L., Haas, Z.J., "Security Ad Hoc Networks," IEEE Network Magazine, Vol. 13, 1999.

[72] Diffie, W., Hellman, M.E., "New Directions in Cryptography," IEEE Transactions on Information Theory, 1976. doi:10.1109/TIT.1976.1055638

[73] Marti, S., Giuli, T.J., Lai, K., Baker, M., "Mitigating Routing Misbehavior in Mobile Ad Hoc Networks," ACM MOBICOM, 2000. doi:10.1145/345910.345955

[74] Perkins, C., Bhagvat, P., "Highly Dynamic Destination-Sequenced Distance Vector Routing for Mobile Computers," ACM SIGCOMM, 1994. doi:10.1145/190809.190336

[75] Basagni, S., Chlamtac, I., Syrotiuk, V., Woodward, B., "A Distance Routing Effect Algorithm for Mobility (DREAM)," ACM MOBICOM 1998. doi:10.1145/288235.288254

[76] Clausen, T., Jacquet, P., Laouti A., Minet, P., Muhlethaler, P., Qayyam, A., Viennot, L., "Optimized Link State Routing Protocol," IETF Internet draft 2001. doi:10.1109/INMIC.2001.995315

[77] Papers from recent proceedings of IEEE Pervasive Computing and Communication 2011 (Percom 2011), March 2011, Seattle, WA.

[78] Vu, L., Do, Q., Nahrstedt, K., "Jyotish: A novel framework for constructing predictive model of people movement from joint Wifi/Bluetooth trace," IEEE PerCom 2011. doi:10.1109/PERCOM.2011.5767595

[79] Zhou, B., Xu, K., Gerla, M., "Group and Swarm Mobility For Ad Hoc Networks Scenarios using Virtual Tracks," IEEE MILCOM 2004.

[80] Pei, G., Gerla, M., Hong, X., Chiang, Ch., "Wireless Hierarchical Routing Protocol with Group Mobility," IEEE Wireless Communications and Networking Conference, 1999. doi:10.1109/WCNC.1999.796996

[81] BS EN ISO 8402, "Quality management and quality assurance – vocabulary," British Standards Institution, 1995.

[82] Leffler, K.B., "Ambiguous changes in product quality," *Am. Rev.*, 1982, 72(5), pp. 956–967.

[83] Garvin, D.A., "What does "product quality" really mean?," *Sloan Management Rev.*, Fall 1984, pp. 25–43.

[84] Crosby, P.B., "Quality without tears, the art of hassle-free management," McGraw Hill, 1984.

[85] Oodan, A.P., Ward, K.E., Mulee, A.W., "Quality of Service in Telecommunications," IEE Telecommunications Series 39, 1997. doi:10.1049/PBTE048E

[86] Lancaster, K.J., "A new approach to consumer theory," *Journal of Political Economy*, April 1996, pp. 132–157. doi:10.1086/259131

[87] Stalling, W., "Wireless Communications & Networks," 2nd Edition, Prentice Hall Pearson, 2005.

[88] Nahrstedt, K., Shah, S., Chen, K., "Cross-Layer Architectures for Bandwidth Management in Wireless Networks," Chapter in 'Resource Management in Wireless Networking,' M. Cardei, I. Cardei, D.Z. Du (Editors), Kluwer Academic Publishers, 2004.

[89] Veres, A., Campbell, A., Barry, M., Sun, L., "Supporting Service Differentiation in Wireless Packet Networks Using Distributed Control," IEEE JSAC (*Journal on Selected Areas*

in Communication), Special Issue on Mobility and Resource Management in Next-Generation Wireless Systems, vol. 19, No. 10, pp. 2094–2104, October 2001. doi:10.1109/49.957321

[90] Xue, Y., Li, B., Nahrstedt, K., "Channel-Relay Price Pair: Towards Arbitrating Incentives in Wireless Ad Hoc Networks," *Journal on Wireless Communications and Mobile Computing*, Special Issue on Ad Hoc Networks, Vol. 6, Issue 2, March 2006. doi:10.1002/wcm.383

[91] Clique-Wikipedia: http://en.wikipedia.org/wiki/Lagrange_multiplier, 2011.

[92] Lagrange-multiplier-Wikipedia: http://en.wikipedia.org/wiki/Lagrange_multiplier, 2011.

[93] Bertsekas, D., Gallager, R., "Data Networks," Prentice Hall 1987.

[94] Nandagopal, T., Kim, T.E., Gao, X., Bhargavan, V., "Achieving MAC Layer Fairness in Wireless Packet Networks," ACM MOBICOM, 2000. doi:10.1145/345910.345925

[95] Tassiulas, L., Sarkar, S., "Maxmin Fair Scheduling in Wireless Networks," IEEE INFOCOM, 2002. doi:10.1109/INFCOM.2002.1019322

[96] Luo, H., Lu, S., Bharghavan, V., "A New Model for Packet Scheduling in Multihop Wireless Networks," ACM MOBICOM, 2000. doi:10.1145/345910.345923

[97] Dharwadker, A., "The Clique Algorithm," The Math Forum Internet Mathematics Library, http://www.mathforum.org/, 2011.

[98] Mulligan, G.D., and Corneil, G., "Corrections to Bierstone's Algorithm for Generating Cliques," Journal of the ACM, Volume 19, Issue 2, April 1972. doi:10.1145/321694.321698

[99] Graph Magics 2.1, Software, http://www.dodownload.com/learning+home+hobby/mathematics/graph+magics.html 2011.

[100] Keshav, S., "An Engineering Approach to Computer Networking," Addison-Wesley Publisher, MA, 1997.

[101] Marsic, I., "Computer Networks: Performance and Quality of Service," December 2004, Department of Electrical and Computer Engineering, Rutgers University, http://www.ece.rutgers.edu/~marsic/Teaching/CCN/minmax-fairsh.html

[102] Perkins, C., and Royer, E., "Ad Hoc On-Demand Distance Vector Routing," 2nd IEEE Workshop on Mobile Computing Systems and Applications (WMCSA), 1999. doi:10.1109/MCSA.1999.749281

[103] Satyanarayanan, M., "Mobile Information Access," IEEE Personal Communications, February 1996. doi:10.1109/98.486973

[104] Noble, B., Satyanarayanan, M., Narayanan, D., Tilton, J.E., Flinn, J., Walker, K.R., "Agile Application-Aware Adaptation for Mobility," ACM Operating Systems Review, 51(5): 276–287, December 1997. doi:10.1145/269005.266708

[105] De Lara, E., Wallach, D.S., Zwaenepoel, W., "Puppeteer: Component-based Adaptation for Mobile Computing," 3rd USENIX Symposium on Internet Technologies and Systems, 2000. doi:10.1145/346152.346224; doi:10.1145/346152.346276

[106] Katz, R.H., "Adaptation and Mobility in Wireless Information Systems," IEEE Personal Communications, 1(1): 6–17, 1994. doi:10.1109/MCOM.2002.1006980

[107] Zheng, R., Kravets, R., "On-demand Power Management for Ad Networks," IEEE INFOCOM 2003. doi:10.1109/INFCOM.2003.1208699

[108] Vardhan, V., et al., "GRACE-2: Integrating fine-grained application adaptation with global adaptation for saving energy," *International Journal of Embedded Systems*, Vol. 4, No. 3, 2009. doi:10.1504/IJES.2009.027939

[109] Yuan, W., Nahrstedt, K., Adve, S., Jones, D.L., Kravets, R., GRACE-1: Cross-Layer Adaptation for Multimedia Quality and Battery Energy," IEEE Transactions on Mobile Computing, 5(7): 799–815, 2006.

[110] Mohapatra, S., Dutt, N., Nicolau, A., Venkatasubramanian, N., "DYNAMO: A Cross-layer Framework for End-to-End QoS and Energy Optimization in Mobile Handheld Devices," *IEEE Journal on Selected Areas in Communications*, 25(4): 722–737, 2007. doi:10.1109/JSAC.2007.070509

[111] http://en.wikipedia.org/wiki/Control_theory, 2011.

[112] Sankar, A., Liu, Z., "Maximum Lifetime Routing in Wireless Ad Hoc Networks," IEEE INFOCOM 2004. doi:10.1109/INFCOM.2004.1356995

[113] Chang, J., Tassiulas, L., "Energy Conserving Routing in Wireless Ad Hoc Networks," IEEE INFOCOM 2000.

[114] Kar, K., Kodialam, M., Lakshman, T.V., Tassiulas, L., "Routing for Network Capacity Maximization in Energy-Constrained Ad Hoc Networks," IEEE INFOCOM 2003. doi:10.1109/INFCOM.2003.1208717

[115] Singh, S., Woo, M., Raghavendra, C., "Power-aware Routing in Mobile Ad Hoc Networks," ACM Mobicom 1998. doi:10.1145/288235.288286

[116] Dasgupta, K., Kukreja, M., Kalpakis, K., "Topology-Aware Placement and Role Assignment for Energy-Efficient Information Gathering in Sensor Networks," IEEE ICC 2003. doi:10.1109/ISCC.2003.1214143

[117] Bhardwaj, M., Chandrakasan, A.P., "Bounding the Lifetime of Sensor Networks Via Optimal Role Assignments," IEEE INFOCOM 2002. doi:10.1109/INFCOM.2002.1019410

[118] Gomez, J., Campbell, A., Naghshineh, M., Bisdikian, C., "Conserving Transmission Power in Wireless Ad Hoc Networks," IEEE ICNP (International Conference on Network Protocols), 2001. doi:10.1109/ICNP.2001.992757

[119] Chang, J., Tassiulas, L., "Fast Approximate Algorithms for Maximum Lifetime Routing in Wireless Ad Hoc Networks," Proceedings of NETWORKING, 2000. doi:10.1007/3-540-45551-5_59

[120] http://people.seas.harvard.edu/~jones/es151/prop_models/propagation.html; 2011.

[121] Linnartz, J.-P., "Wireless Communication: The Interactive Multimedia CD-ROM," Baltzer Science Publishers/Kluwer Academic, Amsterdam, ISSN 1383 4231, 2001.

[122] http://en.wikipedia.org/wiki/IEEE_802.11; http://compnetworking.about.com/od/wireless80211/; http://en.wikipedia.org/wiki/WiMAX; http://en.wikipedia.org/wiki/Open_spectrum; http://en.wikipedia.org/wiki/Bluetooth, July 2011.

[123] Haas, Z., and Pearlman, M.R., "ZRP: A Hybrid framework for routing in Ad Hoc Networks," Ad Hoc Networking, Addison-Wesley Longman Publishing Co., Inc., Boston, MA, USA, 2001.

Author Biography

Klara Nahrstedt is a full professor at the University of Illinois at Urbana-Champaign, Computer Science Department. Her current research interests are directed toward quality of service (QoS) and resource management in wireless and Internet networks, peer-to-peer streaming systems, pervasive computing and communications, and tele-immersive systems. She and her MONET (Multimedia Operating and Networking Systems) research group have worked on many research projects in multimedia distributed systems, multimedia networking, QoS routing, soft-real-time scheduling, mobile systems and networks, multimedia security, privacy and multimedia applications, which led to over 200 refereed papers in journals and conferences. Professor Nahrstedt has also co-authored widely used multimedia books 'Multimedia: Computing, Communications and Applications' published by Prentice Hall, and 'Multimedia Systems' published by Springer Verlag.

Klara Nahrstedt is the recipient of the Early NSF Career Award, Junior Xerox Award, IEEE Communication Society Leonard Abraham Award for Research Achievements, University Scholar Award, Humboldt Research Award, and Ralph and Catherine Fisher Professorship. She is a Fellow of the IEEE.

Klara Nahrstedt received her BA in Mathematics from Humboldt University, Berlin, in 1983, and Diploma degree in numerical analysis from the same university in 1985. She was a research scientist in the Institute for Informatik in Berlin until 1990. In 1995, she received her PhD from the University of Pennsylvania in the Department of Computer and Information Science.

Printed in the United States
by Baker & Taylor Publisher Services